KT-166-064

WONDERS OF THE SOLAR SYSTEM

For Gia, George and Mo, who kept our family together whilst I was away filming Wonders.
Brian Cox

For Anna, Benjamin, Martha and Theo, the true wonders of my solar system.
Andrew Cohen

HarperCollins Publishers
77-85 Fulham Palace Road
London W6 8JB

www.harpercollins.co.uk

Collins is a registered trademark of HarperCollins Publishers Ltd.

First published in 2010

Text © Brian Cox and Andrew Cohen 2010

Photographs, with the exception of those detailed on p254 © BBC

Infographics, Design & Layout © HarperCollins Publishers 2010

By arrangement with the BBC

The BBC logo is a trademark of the British Broadcasting Corporation and is
used under licence.

BBC logo © BBC 1996

15 14 13 12 11 10

9 8 7 6 5 4 3 2 1

The authors assert their moral right to be identified as the authors of this work.

All rights reserved. No parts of this publication may be reproduced, stored
in a retrieval system or transmitted, in any form or by any means, electronic,
mechanical, photocopying, recording or otherwise, without the prior
permission of the publishers.

A catalogue record for this book is available from the British Library.

ISBN 978 0 00 738690 1

Collins uses papers that are natural, renewable and recyclable products
made from wood grown in sustainable forests. The manufacturing processes
conform to the environmental regulations of the country of origin.

All reasonable efforts have been made by the publishers to trace the copyright
owners of the material quoted in this book and of any images reproduced
in this book. In the event that the publishers are notified of any mistakes or
omissions by copyright owners after publication of this book, the publishers
will endeavour to rectify the position accordingly for any subsequent printing.

Associate Publisher: Myles Archibald

Senior Project Editor: Helen Hawksfield
Senior Editor: Julia Koppitz

Cover Design: HarperCollins Publishers

Book interior Design and Art Direction: Studio8 Design

Infographics: Nathalie Lees

Production: Stuart Masheter

Colour reproduction by Saxon Digital Services

Printed in Germany by Mohn media

Mixed Sources
Product group from well-managed
forests and other controlled sources
www.fsc.org Cert no. SW-COC-001806
© 1996 Forest Stewardship Council

FSC is a non-profit international organisation established to promote the
responsible management of the world's forests. Products carrying the FSC
label are independently certified to assure consumers that they come
from forests that are managed to meet the social, economic and
ecological needs of present and future generations.

Find out more about HarperCollins and the environment at
www.harpercollins.co.uk/green

WONDERS OF THE SOLAR SYSTEM

PROFESSOR BRIAN COX
& ANDREW COHEN

Collins

CHAPTER 1
—
THE WONDER
—

CHAPTER 2
—
EMPIRE OF THE SUN
—

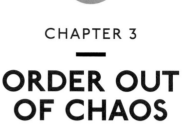

CHAPTER 3
—
ORDER OUT OF CHAOS
—

CHAPTER I

THE WONDER

There are worlds of searing heat and intense cold; planets with winds beyond the harshest of terrestrial hurricanes and moons with great sub-surface oceans of water.

On a frosty winter's afternoon the day after New Year 1959, a tiny metal sphere named *First Cosmic Ship* ascended, gently at first but with increasing ferocity, into a January sky above the Baikonur Cosmodrome a hundred miles east of the Aral Sea. Just a few minutes into her flight the little spacecraft separated from the third stage of her rocket and became the first man-made object to escape the gravitational grasp of planet Earth. On 4 January she passed by the Moon and entered a 450-day long orbit around the Sun somewhere between Earth and Mars, where she remains today. *First Cosmic Ship*, subsequently renamed *Luna 1*, is the oldest artificial planet and mankind's first explorer of the Solar System.

Half a century later, we have launched an armada of robot emissaries and twenty-one human explorers beyond Earth's gravitational grip and onwards to our neighbouring worlds. Five of our little craft have even escaped the gravitational embrace of the Sun on journeys that will ultimately take them to the stars. No longer confined to Earth, we now roam freely throughout the Empire of the Sun.

Our spacecraft have returned travellers' tales and gifts beyond value; an explorer's bounty that easily rivals the treasures, both material and intellectual, from the voyages of Magellan, Drake, Cook and their fellow navigators of Earth's oceans. As with all exploration, the journeys have been costly and difficult, but the rewards are priceless.

Our first half-century of space exploration – less than a human lifetime – has revealed that our Solar System is truly a place of wonders. There are worlds beset with violence and dappled with oases of calm; worlds of fire and ice, searing heat and intense cold; planets with winds beyond the harshest of terrestrial hurricanes and moons with great sub-surface oceans of water. In one corner of the Sun's empire there are planets where lead would flow molten across the surface, in another there are potential habitats for life beyond Earth. There are fountains of ice, volcanic plumes of sulphurous gases rising high into skies bathed in radiation and giant gas worlds ringed with pristine frozen water. A billion tiny worlds of rock and ice orbit our middle-aged yellowing Sun, stretching a quarter of the way to our nearest stellar neighbour, Proxima Centauri. What an empire of riches, and what a subject for a television series.

When we first began discussing making *Wonders of the Solar System*, we quickly realised that there is much

BELOW LEFT: A false colour close-up of the Great Red Spot, originally taken by the *Voyager 1* probe. The spot is a gigantic, stable weather system above the surface of Jupiter.

OVERLEAF: The barred spiral galaxy NGC 1672 in the constellation Dorado, seen by the Hubble Space Telescope.

more to the exploration of space than the spectacular imagery and surprising facts and figures returned by our robot spacefarers. Each mission has contributed new pieces to a subtle and complex jigsaw, and as the voyages have multiplied, a greatly expanded picture of the majestic arena within which we live our lives has been revealed. Mission by mission, piece by piece, we have learnt that our environment does not stop at the top of our atmosphere. The subtle and complex gravitational interactions of the planets with our Sun and the billions of lumps of rock and ice in orbit around it have directly influenced the evolution of the Earth over the 4.5 billion years since its formation, and that influence continues today.

Our moon, unusually large for a satellite in relation to its parent planet, is thought to stabilise our seasons and therefore may have played an important role in allowing complex life on Earth to develop; we probably need our moon.

It is thought that comets from the far reaches of the outer Solar System delivered much of the water in Earth's oceans in a violent bombardment only half a billion years after our home planet formed. This event, known as the Late Heavy Bombardment, is thought to have been the result of an intense gravitational dance between the giant planets of our Solar System, Jupiter, Saturn and Neptune. Without this chance, violent intervention, there may have been little or no water on our planet; we probably need the comets, Jupiter, Saturn and Neptune.

More worryingly for us today, we have no reason to assume that this often-violent interaction with the other inhabitants of the Solar System has ceased; colossal lumps of rock and ice will visit us again from the distant vaults of the outer Solar System, and if undetected and unprepared for, we may not survive. It is thought that this very real threat is diminished by the gravitational influence of Jupiter on passing comets and asteroids, deflecting many of them out of our way; we probably need Jupiter.

This picture of the Solar System as a complex, interwoven and interacting environment stretching way beyond the top of our atmosphere is one of the central stories running throughout the series. In order to understand our place in space we must lift our eyes upwards from Earth's horizon and gaze outwards across a vast sphere extending perhaps for a light year beyond the Kuiper belt of comets and onwards to the edge of the Oort cloud of ice-worlds surrounding the Sun.

Our exploration of the Solar System has also given us very valuable insights into some of the most pressing and urgent problems we face today. Understanding the Earth's complex weather and climate system, and how it responds to changes such as the increase of greenhouse gases in the atmosphere, is perhaps the greatest challenge for early twenty-first century science. This is a planetary science challenge: Earth is one example of a planet with an atmosphere of a particular chemical composition, orbiting the Sun at a particular distance. For a scientist, having only one example of such a vast and interdependent system is less than ideal, and makes the task of understanding its subtle and complex behaviour tremendously difficult. Fortunately, we do have more than one example. Our Solar System is a cosmic laboratory, a diverse collection of hundreds of worlds, large and small, hot and cold. Some orbit closer to our star than we do, most are vastly further away. Some have atmospheres rich in greenhouse gases, far denser than our own, others have lost all but faint traces of their atmospheres to the vacuum of space. Our two planetary neighbours, Venus and Mars, are salient examples of what can happen to worlds very similar to Earth if conditions are slightly different. Venus experienced a runaway greenhouse effect that raised its surface temperature to over 400 degrees Celsius and atmospheric pressure to ninety times that of Earth. Because the laws of physics that control the evolution of planetary atmospheres are the same on Earth and Venus, our understanding of the Greenhouse Effect on Earth can be transferred to and tested on Venus. This provides valuable additional information that can be used to tune and improve those models. The discovery that this benign, blue planet which shimmers brightly and with such beauty in the twilight skies of Earth was transformed long ago into a hellish world of searing temperatures, acid rain and crushing pressure has had a genuine and profound psychological effect because it demonstrates in stark terms that runaway greenhouse effects can happen to planets not too dissimilar from our own.

Similar salutary insights have accrued from our studies of Mars. In the early days of Mars exploration, observations of the evolution of dust storms on the red planet provided support for the nuclear winter hypothesis on Earth. Storms that began in small local areas were observed to throw large amounts of dust into the Martian atmosphere. Over a period of a few weeks, the dust encircled the entire planet,

Reaching for worlds beyond our grasp is an essential driver of progress and necessary sustenance for the human spirit. Curiosity is the rocket fuel that powers our civilization.

exactly matching the models of the evolution of the dust and smoke clouds that would be created by a large-scale nuclear exchange. When a planet is shrouded in dust, the warmth of the Sun is reflected back into space and temperatures quickly fall, leading to a so-called nuclear winter that could last many decades. On Earth, this could lead to the extinction of many species, including perhaps our own.

The observation of a mini-nuclear winter periodically playing itself out on Mars was a major factor in the acceptance of the theory, and this in turn profoundly influenced the thinking of major players at the end of the cold war. As former Russian president Mikhail Gorbachev said in 2000, 'Models made by Russian and American scientists showed that a nuclear war would result in a nuclear winter that would be extremely destructive to all life on Earth; the knowledge of that was a great stimulus to us, to people of honor and morality, to act in that situation'.

Wonders is also a story of human ingenuity and engineering excellence. Russia's *First Cosmic Ship* began its voyage beyond Earth just fifty-five years after the first powered flight by Orville and Wilbur Wright in December 1903. *Wright Flyer 1* was constructed from spruce and muslin, and powered by a twelve-horsepower petrol engine assembled in a bicycle repair shop. By 1969, less than one human lifetime away, Armstrong and Aldrin set foot on another world, launched by a Saturn V rocket whose giant

BELOW: Astronaut Ed White made the first American spacewalk during the *Gemini 4* mission on 3 June 1965, over the Pacific Ocean for twenty-three minutes.

OVERLEAF: Lift-off of the *Apollo 17 Saturn V Moon Rocket* from Kennedy Space Centre, Florida on 17 December 1972.

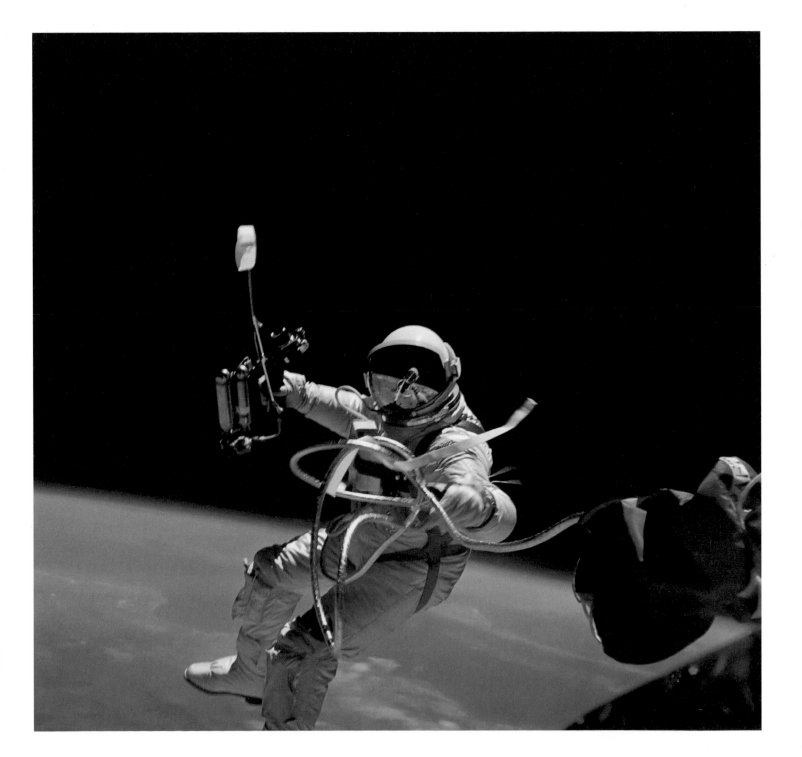

14

THE WONDER

first-stage engines generated around 180 million horsepower between them. The most powerful and evocative flying machine ever built, the Moon rocket stood 111 metres (364 feet) high, just thirty centimetres (twelve inches) short of the dome of Wren's magisterial St Paul's Cathedral. Fully fuelled for a lunar voyage, it weighed 3,000 tonnes. Sixty-six years before *Apollo 11*'s half-a-million mile round trip to the Moon, *Wright Flyer 1* reached an altitude of three metres (ten feet) on its maiden voyage. This rate of technological advancement, culminating with our first journeys into the deep Solar System, is surely unparalleled in human history, and the benefits are practically incalculable.

Most importantly of all, *Wonders* is a celebration of the spirit of exploration. This is desperately relevant, an idea so important that celebration is perhaps too weak a word. It is a plea for the spirit of the navigators of the seas and the pioneers of aviation and spaceflight to be restored and cherished; a case made to the viewer and reader that reaching for worlds beyond our grasp is an essential driver of progress and necessary sustenance for the human spirit. Curiosity is the rocket fuel that powers our civilization. If we deny this innate and powerful urge, perhaps because earthly concerns seem more worthy or pressing, then the borders of our intellectual and physical domain will shrink with our ambitions. We are part of a much wider ecosystem, and our prosperity and even long-term survival are contingent on our understanding of it.

In 1962, John F. Kennedy made one of the great political speeches at Rice University in Houston, Texas. In the speech, he argued the case for America's costly and wildly ambitious conquest of the Moon. Imagine the bravado, the sheer power and confidence of vision in committing to a journey across a quarter of a million miles of space, landing on another world and returning safely to Earth. It might have been perceived as hubris at the time, but it worked. America achieved this most audacious feat of human ingenuity within nine years of launching their first manned sub-orbital flight. Next time you glance up at our shimmering satellite, give a thought to the human beings just like you who decided to go there and plant their flag for all mankind.

At the turn of the twenty-first century, the Solar System is our civilization's frontier. Our first steps into the unexplored lands above our heads have been wildly successful, revealing a treasure chest of new worlds and giving us priceless insights into our planet's unique beauty and fragility. As a species, we are constantly balanced on a knife-edge, prone to parochial disputes and unable to harness our powerful curiosity and boundless ingenuity. The exploration of the Solar System has brought out the best in us, and this is a precious gift. It rips away our worst instincts and forcibly thrusts us into a face-to-face encounter with the best. We are compelled to understand that we are one species amongst millions, living on one planet around one star amongst billions, inside one galaxy amongst trillions. The beauty of our planet is made manifest, enhanced immeasurably by the juxtaposition with other worlds. Ultimately, the value of young, curious and wonderful humanity as we take our first steps outwards from our home world is brought into such startling relief that all who share in the wonder must surely be filled with optimism and a powerful desire to continue this most valuable of journeys ◉

'Many years ago the great British explorer George Mallory, who was to die on Mount Everest, was asked why did he want to climb it. He said, "Because it is there." Well, space is there, and we're going to climb it, and the Moon and the planets are there, and new hopes for knowledge and peace are there. And, therefore, as we set sail we ask God's blessing on the most hazardous and dangerous and greatest adventure on which man has ever embarked.'

— John F. Kennedy, Rice University 1962

RIGHT: The first manned lunar landing mission, *Apollo 11*, launched from the Kennedy Space Centre on 16 July 1969. Aboard the spacecraft were astronauts Neil Armstrong, Michael Collins and Edwin E. (Buzz) Aldrin. In this photograph, Aldrin walks past some rocks, easily carrying scientific equipment which would have been too heavy to carry on Earth.

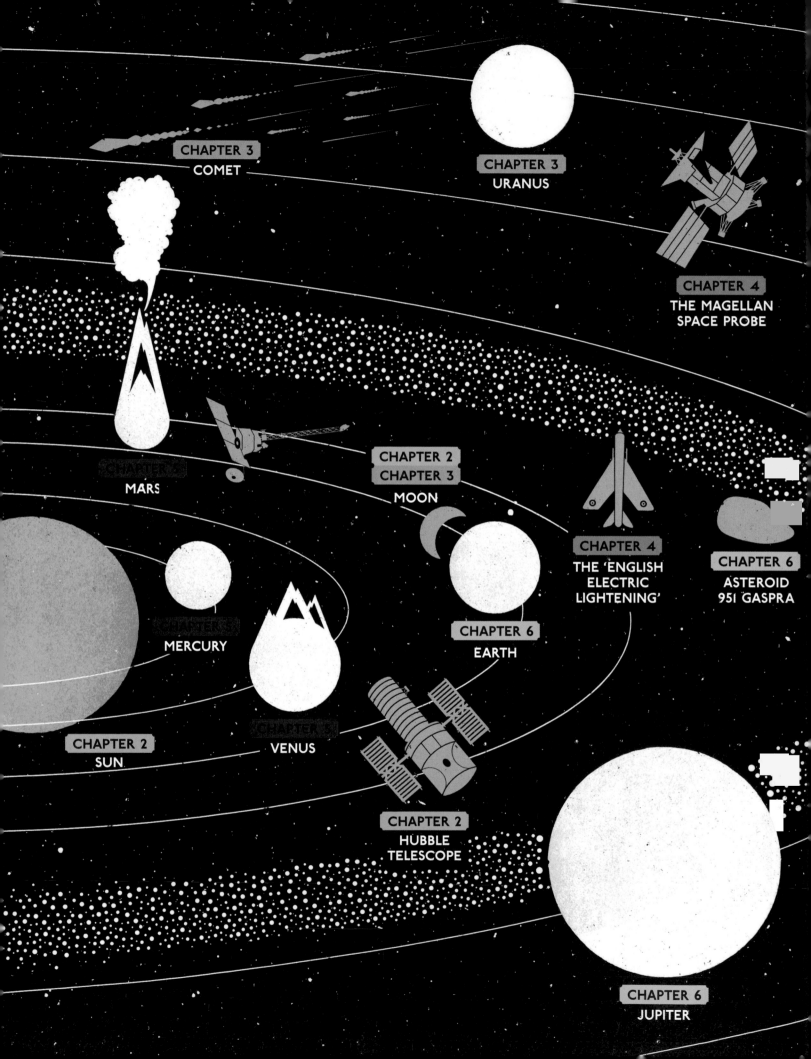

CHAPTER 3
COMET

CHAPTER 3
URANUS

CHAPTER 4
THE MAGELLAN
SPACE PROBE

CHAPTER 5
MARS

CHAPTER 2
CHAPTER 3
MOON

CHAPTER 4
THE 'ENGLISH
ELECTRIC
LIGHTENING'

CHAPTER 6
ASTEROID
951 GASPRA

CHAPTER 5
MERCURY

CHAPTER 6
EARTH

CHAPTER 2
SUN

CHAPTER 5
VENUS

CHAPTER 2
HUBBLE
TELESCOPE

CHAPTER 6
JUPITER

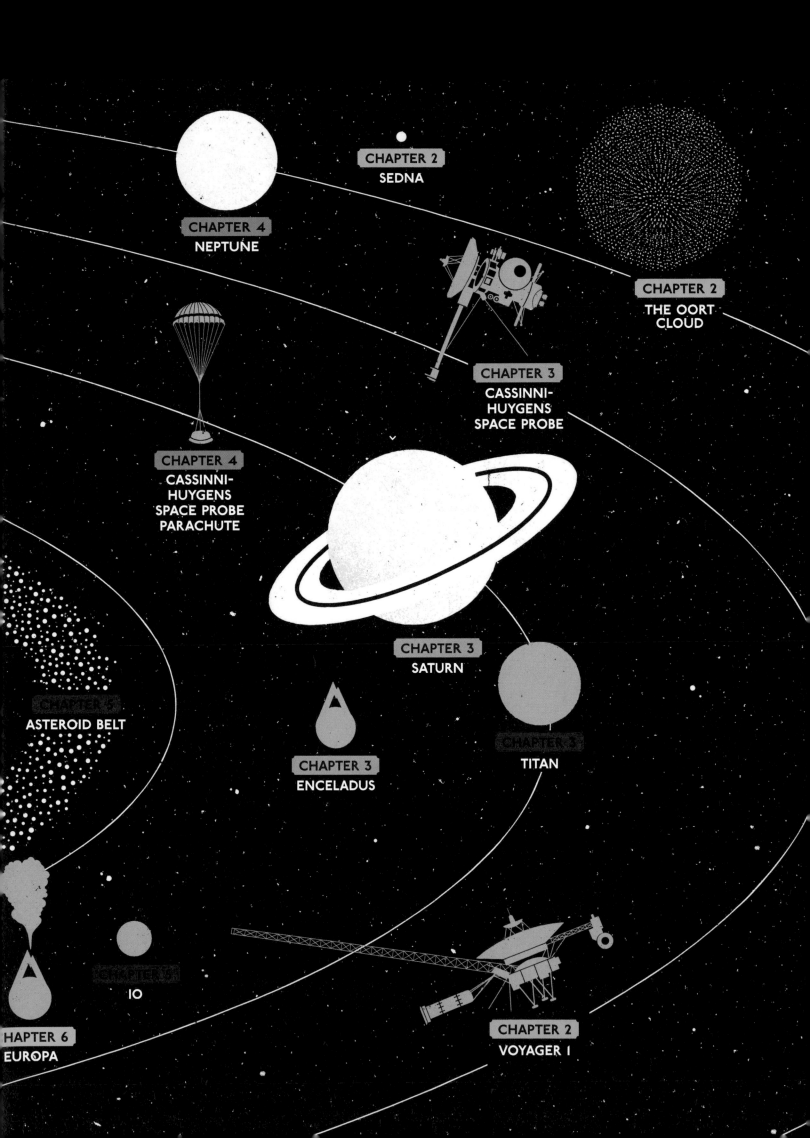

CHAPTER 2
SEDNA

CHAPTER 4
NEPTUNE

CHAPTER 2
THE OORT
CLOUD

CHAPTER 3
CASSINNI-
HUYGENS
SPACE PROBE

CHAPTER 4
CASSINNI-
HUYGENS
SPACE PROBE
PARACHUTE

CHAPTER 3
SATURN

CHAPTER 5
ASTEROID BELT

CHAPTER 3
TITAN

CHAPTER 3
ENCELADUS

CHAPTER 6
IO

CHAPTER 6
EUROPA

CHAPTER 2
VOYAGER I

CHAPTER 2

EMPIRE OF THE SUN

AN ORDINARY STAR

At the heart of our complex and fascinating Solar System sits its powerhouse. For us it is everything, and yet it is just one ordinary star amongst 200 billion stars within our galaxy. It is a large wonder that greets us every morning; a star that controls each and every world that it holds in its thrall – the Sun. The Sun reigns over a vast empire of worlds and without it we would be nothing; life on Earth would not exist. Although we live in the wonderous empire of the Sun, it is a place we can never hope to visit. However, thanks to the continual advances in technology and space exploration, and through observation from here on Earth, each spectacular detail we see leads us closer to understanding the enigma that is the Sun.

BELOW: In July 2009, pilgrims flocked to Varanasi, on the sacred river Ganges, not just to bathe in the holy waters but also to watch the longest total eclipse of the Sun of the twenty-first century.

In the north of India, on the banks of the river Ganges, lies the holy city of Varanasi. It is one of the oldest continuously inhabited cities in the world, and for Hindus it is one of the holiest sites in all of India.

Varanasi, or Banares, to give the city its old name, is a city suffused with the colours, sounds and smells of a more ancient India. Mark Twain famously wrote: 'Banares is older than history, older than tradition, older even than legend, and looks twice as old as all of them put together'.

Each year a million pilgrims visit Varanasi to bathe in the holy river and pray in the hundreds of temples that cover the city. Part of what makes the city so special is the orientation of its sacred river as it flows past; it's the only place where the Ganges turns around to the north, making it the one spot on the river where you can bathe while watching the Sun rise on the eastern shore. And sunrise over Varanasi is certainly one of Earth's great sights. The humid, tropical air adds a soporific quality to the light, which in turn lends a fairy-tale quality to the brightly coloured buildings and palaces that line the holy river. It is a misty, pastel-shaded, dream-like experience, as though the city is materialising not from the dawn but from the past.

But on 22 July 2009, at precisely 6.24am, a different type of pilgrim was to be found waiting beside the Ganges to witness one of the true wonders of the Solar System.

At this time, across a small strip of the Earth's surface, the longest total eclipse of the Sun since June 1991 was about to be visible to a lucky few. For three and a half minutes the Moon would cover the face of the Sun and plunge this ancient city into darkness ◉

SOLAR ECLIPSES

A total solar eclipse is possibly the most visual and visceral example of the structure and rhythm of our solar system. It is a very human experience, and one that lays bare the mechanics of this system.

At the centre is the Sun, reigning over an empire of worlds that move like clockwork. Everything within its realm obeys the laws of celestial mechanics discovered by Sir Isaac Newton in the late seventeenth century. These laws allow us to predict exactly where every world will be for centuries to come. And wherever you happen to be, if there's a moon between you and the Sun, there will be a solar eclipse at some point in time.

Eclipses occur all over the Solar System; Jupiter, Saturn, Uranus and Neptune all have moons and so eclipses around these planets are a frequent occurrence. On Saturn, the moon Titan passes between the Sun and the ringed planet every fifteen years, while on the planetoid Pluto, eclipses with its large moon, Charon, occur in bursts every 120 years. But the king of eclipses is the gas giant Jupiter; with four large moons orbiting the planet, it's common to see the shadow of moons such as Io, Ganymede and Europa moving across the Jovian cloud tops. Occasionally the eclipses can be even more spectacular. In spring 2004, the Hubble telescope took a rare picture (opposite top) in which you can see the shadows of three moons on Jupiter's surface; three eclipses occurring simultaneously. Although this kind of event happens only once every few decades, the timing of Jupiter's eclipses is as predictable as every other celestial event. For hundreds of years we've been able to look up at the night sky and know exactly what will happen when. Historically, this precise understanding of the motion of the Solar System provided the foundation upon which a much deeper understanding of the structure and workings of our universe rests. A wonderful example is the extraordinary calculation performed by the little-known

Dutch astronomer Ole Romer in the 1670s. Romer was one of many astronomers who attempted to solve a puzzle that seemed to make no sense.

The eclipses of the Galilean moons, Io, Europa, Ganymede and Callisto, by Jupiter were accurately predicted once their orbits had been plotted and understood. But it was soon observed that the moons vanished and reappeared behind Jupiter's disc about twenty minutes later than expected when Jupiter was on the far side of the Sun (the accurate modern figure is seventeen minutes). When the predictions of a scientific theory disagree with evidence the theory must be modified or even rejected, unless an explanation can be found. Newton's beautiful clockwork Solar System was on trail.

Romer was the first to realise that this delay was not a glitch in the clockwork Solar System. Instead, it was caused because light takes time to travel from Jupiter to Earth. The eclipses of the Galilean moons happen just as Newton predicts, but we don't see the eclipses on earth until slightly later than predicted when Jupiter moves further away from the Earth, simply because it takes more time for light to travel a greater distance.

From this beautifully simple observation of the eclipses of Jupiter, fellow Dutch astronomer, Christian Huygens, was able to make the first calculation of the speed of light. The speed of light, we now know, is a fundamental property of our universe. It is one of the universal numbers that is unchanging and fixed throughout the cosmos. Ultimately, an understanding of its true significance had to wait until Einstein's theory of space and time, Special Relativity, in 1905, but the long and winding road of discovery can be traced back to Romer and his eclipses.

Closer to home, eclipses become even more familiar. In 2004, the Mars Exploration Rover Opportunity looked up from the surface of Mars and took possibly the most beautiful picture of any extraterrestrial eclipse (opposite bottom). In this remarkable image you can see Mars' moon Phobos as it makes its way across the Sun – an image of a partial solar eclipse from the surface of another world.

Eclipses on Mars are not only possible but commonplace, with hundreds occurring each year, but

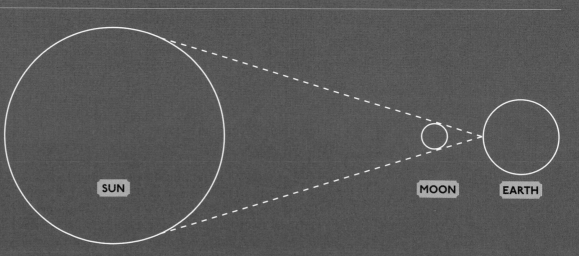

A TOTAL SOLAR ECLIPSE
This natural phenomenon occurs when the moon passes between the Earth and the Sun, completely obscuring the view of the Sun from Earth.

SUN

MOON   EARTH

one event we will never see on Mars is a total solar eclipse. Here on Earth, though, humans have the best seat in the Solar System from which to enjoy the spectacle of a total eclipse of the Sun – all thanks to a wonderful quirk of fate.

For a perfect total solar eclipse to occur, a moon must appear to be exactly the same size in the sky as the Sun. On every other planet in the Solar System the moons are the wrong size and the wrong distance from the Sun to create the perfect perspective of a total solar eclipse. However, here on Earth the heavens have arranged themselves in perfect order. The Sun is 400 times the diameter of the Moon and, by sheer coincidence, it's also 400 times further away from the Earth. So when our moon passes in front of the Sun it can completely obscure it.

With over 150 moons in the Solar System you might expect to find other total solar eclipses, but none produce such perfect eclipses as the Earth's moon. It won't last forever, though; The clockwork of the Solar System is such that the raising of the tides on Earth caused by the Moon has consequences. As the Earth spins beneath tidal bulges raised by the Moon, its rate of rotation is gradually, almost imperceptibly, reduced by friction, and this has the effect of causing the Moon to gradually drift further and further away from Earth. This complex dance, in precise accord with Newton's laws, is also responsible for the fact that we only see one side of the Moon from the Earth – a phenomenon called spin-orbit locking.

The drift is tiny, only around 4 centimetres (1.6 inches) per year, but over the vast expanses of geological time it all adds up. Around 65 million years ago the Moon was much closer to Earth and the dinosaurs would not have been able to see the perfect eclipses we see today. The Moon would have been closer to Earth and would therefore have completely blotted out the Sun with room to spare. In the future, as the Moon moves away from the Earth, the unique alignment will slowly begin to degrade; while drifting away from our planet the Moon will become smaller in the sky and eventually too small to cover the Sun. This accidental arrangement of the Solar System means that we are now living in exactly the right place and at exactly the right time to enjoy the most precious of astronomical events ◉

BELOW: The king of eclipses. This image from NASA's Hubble space telescope clearly shows three black circles on Jupiter's surface, which are the shadows cast by three of the four moons that circle the planet.

BOTTOM: The three images from left to right are taken by the Mars Exploration Rover Opportunity and show the journey of Mars's moon, Phobos, as it passes over the disk of the Sun in a partial solar eclipse.

IN THE REALM
OF THE SUN

Our closest star is the strangest, most alien place in the Solar System. It's a place we can never hope to visit, but through space exploration and a few chance discoveries our generation is getting to know the Sun in exquisite new detail. For us it's everything, and yet it's just one ordinary star among 200 billion starry wonders that make up our galaxy. To explore the realm of our sun requires a journey of over thirteen billion kilometres; a journey that takes us from temperatures reaching fifteen million degrees Celsius, in the heart of our star, to the frozen edge of the Solar System where the Sun's warmth has long disappeared.

On 14 November 2003, three American scientists discovered a dwarf planet at the remotest frontier of the Solar System. Sedna is a planetoid three times more distant from the Sun than Neptune. Around 1,600 kilometres (1,000 miles) in diameter, Sedna is barely touched by the Sun's warmth; its surface temperature never rises above minus -240 degrees Celsius. For most of its orbit Sedna is further from our star than any other known dwarf planet. On its slow journey around the Sun, one complete orbit – Sedna's year – takes 12,000 Earth years. From its frozen surface at least thirteen billion kilometres from Earth, a view of the Sun rising on Sedna would give a very different perspective on our solar system and a clear depiction of how far the Sun's realm stretches. Sunrise on Sedna is no more than the rising of a star in the night sky: from this frozen place, our blazing sun is just another star.

To travel from the outer reaches of Sedna's orbit to one of the first true planets of the Solar System we would need to cover over ten billion kilometres. Uranus was the first planet to be discovered with the use of a telescope, in 1781, by Sir William Herschel, and like all the giant planets (except Neptune) it is visible with the naked eye. Even so, sunrise on Uranus is barely perceptible; the Sun hangs in the sky 300

SUN

0.15 BILLION KM   EARTH

LEFT: Newly discovered dwarf planet Sedna is located at the most remote frontier of the Solar System. From this frozen planet the Sun would appear merely as a remote star.

BELOW: This atmospheric shot of the Sun setting behind the Gusev crater on Mars was captured by the Exploration Rover Spirit. Sunrise and sunset can each take two hours.

times smaller than it appears on Earth. Only when we have travelled the two and a half billion kilometres past Jupiter and Saturn do we arrive at the first world with a more familiar view of the Sun. Over 200 million kilometres out, sunset on Mars is a strangely familiar sight. On 19 May 2005, the Mars Exploration Rover Spirit captured this eerie view as the Sun sank below the rim of the Gusev crater. The panoramic mosaic image was taken by the rover at 6.07pm, on the rover's 489th day of residency on the red planet. This sunset shot is not only beautiful, but it also tells us something fundamental about the Martian sky. Repeated observations have revealed that twilight on Mars is a rather long affair, lasting for up to two hours before sunrise and two hours after sunset. The reason for this long slow progression to and from darkness is the fine dust that is whipped up off the surface of Mars and lifted to incredibly high altitudes. At this height the Sun's rays are scattered by the dust from the sunlit side of Mars around to the dark side, producing the long, leisurely and beautiful journey between day and night. Here on Earth, some of the longest and most spectacular sunrises and sunsets are produced by a similar mechanism, when tiny dust grains are catapulted high into the atmosphere by powerful volcanoes, scattering light into extra-colourful moments on our planet.

Moving past Earth, which is 150 million kilometres out, we head to the heart of the Solar System. Mercury is the closest planet to the Sun, just forty-six million kilometres (twenty-nine million miles) away. It spins so slowly that sunrise to sunrise lasts for 176 Earth days. Beyond it there is nothing but the naked Sun, a colossal fiery sphere of tortured matter burning with a core temperature of about fifteen million degrees Celsius. The sheer scale of the Sun is difficult to conceive; at 140,000 kilometres (865,000 miles) across it is over 100 times the diameter of Earth, which means you could fill it with over a million Earths. Its mass is 2×10^{30} kilograms – 330,000 times that of our planet. If you add up the masses of all the planets, dwarf planets, moons and asteroids, you would find they contribute less than half a per cent of the total mass of the Solar System. The Sun is dominant – the rest is an afterthought ◉

13 BILLION KM

SEDNA

Throughout human history, this majestic wonder has been a constant source of comfort, awe and worship, but our understanding of the Sun has developed slowly. For centuries, the finest minds in science struggled to understand how it created such a seemingly endless source of heat and energy. As recently as the nineteenth century science had little knowledge of what the Sun was made of, where it had come from, or the secret of its phenomenal power.

LEFT: Our brightest star. The Sun is just one of billions of stars, but to all living things on Earth it is of crucial importance – without it no life could exist.

THE ENERGY
OF THE SUN

BELOW LEFT: Using just an umbrella, a tin of water and a thermometer, we measured the energy given off by the Sun in Death Valley, California – regularly the hottest place on the planet.

RIGHT: What appears to be a hole in the night sky is actually Molecular Cloud Barnard 68. This cloud of dust and molecular gas will eventually form a bright new star system.

In 1833, John Herschel, the most famous astronomer of his generation, travelled to the Cape of Good Hope in South Africa on an ambitious astronomical adventure to map the stars of the southern skies. This voyage was the end of an extraordinary odyssey for the Herschel family; he completed the work his father, William Herschel, had begun in the northern skies 50 years earlier.

In 1838, Herschel attempted to answer one of the most fundamental questions we can ask about the Sun – how much energy does it produce? It may seem an incredibly ambitious calculation, but Herschel knew that to measure this 'solar constant' he would need nothing more than a thermometer, a tin of water, an umbrella and the predictable blue skies of Cape Town.

When you want to measure the Sun's radiation across billions of miles of space you need to start small. So Herschel began by asking how much energy the Sun delivers onto a small part of the Earth's surface – in this case, onto a tin full of water. Herschel waited until December to conduct this experiment, when the Sun would be directly overhead, then placed his tin under the shade of the umbrella in the midday Sun. Once the water had heated up to ambient temperature he removed the shade to allow the Sun to shine directly onto the water. In direct sunlight, the water temperature begins to rise and by timing how long it takes the Sun to raise the water temperature by one degree Celsius, Herschel could calculate exactly how much energy the Sun delivered into the can of water.

The calculation was simple because Herschel already knew something called the specific heat capacity of water – in modern units it is the amount of energy required to raise the temperature of 1 kilogramme of water by one Kelvin. Kelvin is a temperature scale usually favoured in science: 1 Kelvin = 1 degree Celsius, and -273 K = 0 degrees Celsius. (For the record, the specific heat capacity of water is 4187 Joules per kilogramme per Kelvin.) From this calculation it's a small step to scale the number up and work out how much energy is delivered to a square metre of the surface of the Earth in one second. It turns out that on a clear day, when the Sun is vertically overhead, that number is about a kilowatt. That equates to ten 100-watt bulbs being powered by the Sun's energy for every metre squared of the Earth's surface.

With this number Herschel could now take a leap of imagination and calculate the entire energy output of the Sun. He knew that the Earth is 150 million kilometres (93 million miles) away from the Sun, so he created an imaginary giant sphere around the Sun with a radius of 150 million kilometres.

Every second the Sun produces 400 million million million million watts of power – that is a million times the power consumption of the United States every year – radiated in one second.

By adding up each of those kilowatts for every square metre of this entirely imaginary sphere, he was able to estimate the total energy output of the Sun per second. It's a number that begins to reveal the sheer magnitude of our star. Every second the Sun produces 400 million million million million watts of power – that is a million times the power consumption of the United States every year – radiated in one second. It's an ungraspable power, but a power that we have calculated using the very simplest of experiments and some water, a thermometer, a tin and an umbrella.

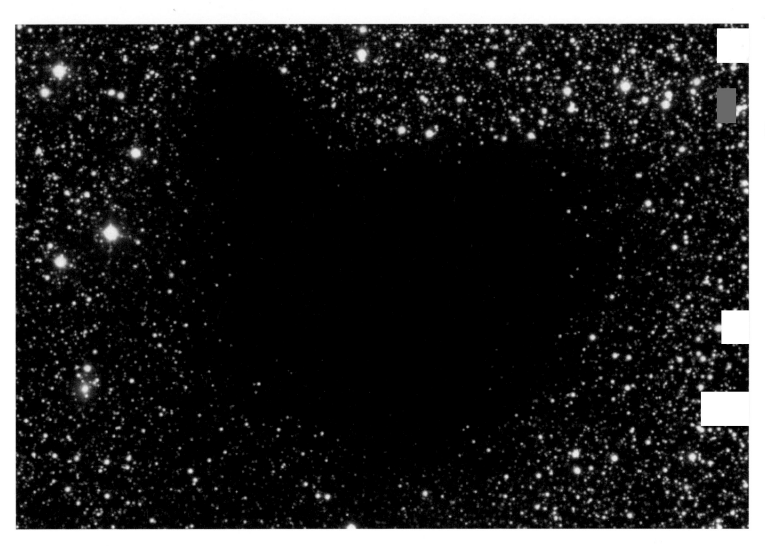

A STAR IS BORN

It's a wonder of the Sun that it has managed to keep up this phenomenal rate of energy production for millennia. Stars like the Sun are incredibly long-lived and stable – our best estimate for the age of the Universe is 13.7 billion years, and the Sun has been around for nearly five billion years of that, making it about a third of the age of the Universe itself. So what possible power source could allow the Sun to shine with such intensity day after day for five billion years? The best way to find the answer is to go back to the beginning, to a time when this corner of the galaxy was without light, and the Sun had yet to begin.

The picture above shows the Milky Way. The dark areas with an absence of stars are called molecular clouds; clouds of molecular hydrogen and dust that are lying between us and the stars of the Milky Way galaxy. Taken by the Very Large Telescope (VLT) at Paranal Observatory, in Chile, this image is of Barnard 68, a molecular cloud well within our galaxy at a distance of about 410 light years. Take a close look, because you are looking at a future star, a cloud of dust and gas that in the next 100,000 years or so will collapse and begin its journey to becoming a new light in the heavens.

Barnard 68, like all molecular clouds, contains the raw material from which stars are made, vast stellar nurseries that are among the coldest, most isolated places in the galaxy. This particular cloud is around half a light year across, or twenty million million kilometres, and has a mass that is about twice that of our sun. Most importantly, it is incredibly cold; in the heart of this cloud the temperature is no more than 4 Kelvin, that's -269 degrees Celsius. That matters because temperature is a measure of how fast things are moving, so in these clouds the clumps of hydrogen and dust are moving very slowly.

The stability of a cloud like Barnard 68 is in a fine balance. On one hand, the clumps of hygrogen and dust are moving around, which leads to an outward pressure that acts to expand the cloud. Counteracting this is the force of gravity – an attractive force between all the particles in the cloud that tries to collapse it inwards. In order for the cloud to become a star, gravity must gain the upper hand long enough to cause a dramatic collapse of the cloud. This can only happen if the particles are moving very slowly, i.e., if the temperature is low.

Over millennia gravity's weak influence dominates and the molecular clouds begin to collapse, forcing the hydrogen and dust together in ever-denser clumps. We have a name for clumps of gas and dust collapsing under their own gravity: stars. As the clouds collapse further and further they begin to heat up and eventually in their cores they become hot enough for the hydrogen to begin to fuse into helium. The stars ignite, the clouds are no longer black and the life cycle of a new star has begun ◉

Five billion years ago a star was born that would come to be known as the Sun. Its birth reveals the secret of our star's extraordinary resources of energy, because the Sun, like every other star, was set alight by the most powerful known force in the Universe.

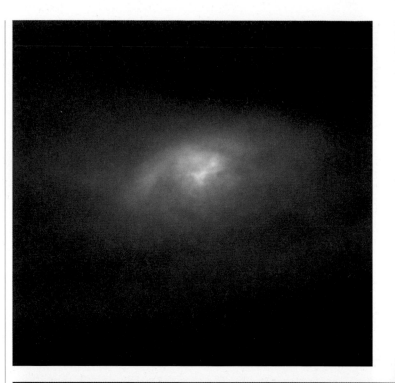

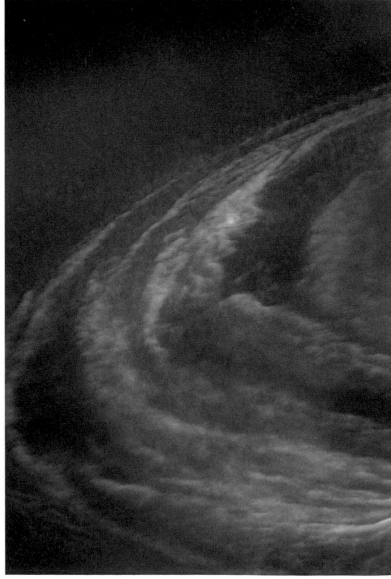

RIGHT: The Sun, as with any other star, was born from giant clouds of molecular dust and gas; it developed into a spinning ball of hot gas that is heated by a thermonuclear reaction at its core.

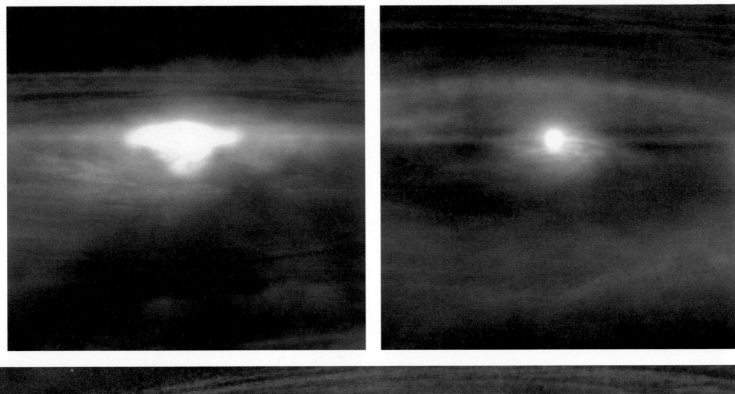

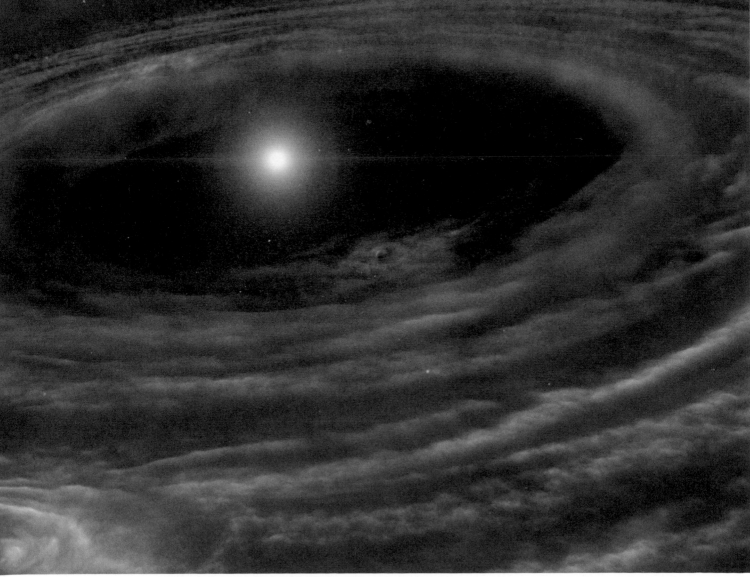

THE FORCES BEHIND THE SUN

RIGHT: The burning handle on the bottom right of the Sun shown here is known as the February 12 'prominence'. Prominences are loops of gas that are suspended above the surface of the Sun, which are anchored in place by opposing magnetic fields.

Nuclear fusion is the process by which all the chemical elements in the Universe, other than hydrogen, were produced. There are just three fundamental building blocks of matter required to make up everything we can see – from the most distant stars to the smallest piece of dust in our Solar System. Two kinds, the Up and Down quarks, make up the protons and neutrons in the atomic nuclei, and a third, the electrons, orbit around the nuclei to make atoms. These particles make up literally everything, including the book you are reading, the hand holding the book and the eyes reading the print. We live in a universe that is simple at heart!

The Universe today is, of course, far from simple. It is a complex, beautiful and diverse place with stars, planets and humans. Nuclear fusion is one of the primary processes that built that complexity.

The Universe began 13.7 billion years ago in the Big Bang. In the first instant it was unimaginably hot and dense, but it expanded and cooled very quickly. After just one second it was cold enough for the Up and Down quarks to stick together into protons and neutrons. The hydrogen nucleus is the simplest in nature, consisting of a single proton. Helium is the next simplest, built of two protons and one or two neutrons. Then comes lithium, beryllium, boron, carbon, nitrogen, oxygen and so on, each with one more proton and accompanying neutrons. This process of sticking more and more protons and neutrons together to form the chemical elements is known as nuclear fusion.

The process of fusion is not easy. Protons carry positive electric charge, which means that they feel a powerful repulsive force when they get close to one another. The force that drives them apart is one of the four fundamental forces of nature: electromagnetism. If the protons can get close enough, another force – called the strong nuclear force – takes over. The strong force is aptly named (it is the strongest in the Universe) and can easily overcome the weaker electromagnetic repulsion. We don't notice the strong force in everyday life because its effects are felt over a very short range and it stays trapped and hidden within the atomic nucleus.

The way to get protons close enough for fusion to occur is to heat them up to very high temperatures. As I've explained before, this is because temperature is a measure of how fast things are moving around; if the protons approach each other at high speed they can overcome the electromagnetic repulsion and get close enough for the strong force to take over and bind them together.

For the first few moments in the life of the Universe, all of space was filled with particles that were hot enough to smash together and fuse, but this only lasted a few brief minutes. Around ten minutes after the Big Bang, the Universe had cooled down enough for fusion to cease. At that time, our Cosmos was approximately 75 per cent hydrogen and 25 per cent helium, with very small traces of lithium. Fusion did not reappear in the Universe until the first stars were born, a few hundred million years later.

The high temperatures inside stars like our Sun mean that the hydrogen nuclei in their cores are moving fast enough for the electromagnetic repulsion to be overcome and the strong nuclear force to take over, initiating nuclear fusion. The process is quite complex and involved, and very, very slow. First, two protons must approach each

NUCLEAR FUSION
Naturally occurring in stars, this is the process by which several atomic nuclei fuse together to create one single heavier nucleus.

● PROTON   ⊕ POSITRON
○ NEUTRON   γ GAMMA RADIATION

⊕ & γ & **ENERGY**   γ & **ENERGY**   **ENERGY**

other to within a thousand million millionth's of a metre (written as 10^{-15} m). Then something very rare must happen – a proton must change into a neutron. This happens through the action of the third of the four forces of nature: the Weak Nuclear Force. The Weak Force is, as its name suggests, unlikely to act: an average proton will live for billions of years in the Sun's core before fusion begins.

When this first step towards fusion finally occurs, a closely bound proton and neutron are formed. This nucleus is known as Deuterium. In the process, an anti-matter electron (known as a positron) and a sub-atomic particle called a neutrino are released. There is also an important extra ingredient, which is the key to understanding why stars shine. If you add up the mass of the Deuterium, the electron and the little neutrino, you find that it is slightly less than the mass of the original two protons. Mass is lost in the fusion process and turned into energy. This is an application of Einstein's most famous equation: $E=mc^2$.

This energy emerges from the Sun as sunshine – it is the primary power source for all life on Earth.

The fusion process then proceeds much more quickly because the action of the Weak Nuclear Force is no longer required. The positron bumps into an electron and disappears in another flash of energy. A proton fuses with the Deuterium nucleus to make a form of helium known as helium 3 (two protons and one neutron), and then two helium 3 nuclei fuse together to form helium 4 – the end product of fusion in the Sun – releasing two protons. At each stage mass is converted to energy, keeping the Sun hot and shining brightly.

At the end of their life, stars run out of hydrogen fuel in their cores and more complex fusion reactions occur. Heavier elements are produced – oxygen, carbon, nitrogen – the elements of life. Every element in the Universe today was fused together from the primordial hydrogen and helium left over from the Big Bang ◉

THE POWER OF SUNLIGHT

Once photons leave the Sun, the journey to Earth is a relatively short one. Light, like all forms of electromagnetic waves, travels at the same speed – almost 300 thousand kilometres a second, and so a photon leaving the surface of the Sun will reach the Earth in about eight minutes. Having travelled almost 150 million kilometres across space, each and every photon has a remarkable ability to shape and transform our planet.

ABOVE: The Iguaçu Falls, located on the border of Brazil and Argentina, is another example of how the power of the Sun shapes the contours of the Earth.

On the border of the Brazilian state of Paraná and the Argentine province of Misiones is the Iguaçu river. Stretching for over one thousand kilometres, the Iguaçu eventually flows into the Parana, one of the great rivers of the world. It's these river systems that eventually drain all the rainfall from the southern Amazonian basin into the Atlantic. Billions of gallons of water flow through this river system each day and all of it, every molecule in the river, every molecule in every raindrop in every cloud, has been transported from the Pacific over the Andes and into the continental interior here by the energy carried in single photons from our sun. The Sun is the power that lifts all the water on the Blue Planet, shaping and carving our landscape and creating some of the most breathtaking sights on Earth.

The Iguaçu Falls are one of the most spectacular natural wonders on our planet. Almost three kilometres (two miles) long, comprising over 275 individual falls and reaching heights of over 76 metres (250 feet), a quarter of a million gallons of water flow through the Falls every second.

The spectacular energy of these waterfalls is a wonderful example of how this planet is hardwired to the seemingly constant and unfailing power of the Sun. For centuries it was assumed that the Sun, like all the heavens, was perfect and unchanging, but gradually we've come to realise that the Sun is far more dynamic then just a perfect beautiful orb in the sky. Even tiny fluctuations in its brightness can have huge effects here on Earth ◉

SUNSPOTS: THE SEASONS OF THE SUN

BOTTOM: This cross-section of a sunspot visualises the temperatures around it. The red areas show where the temperature is higher than average; the blue where it is lower than average.

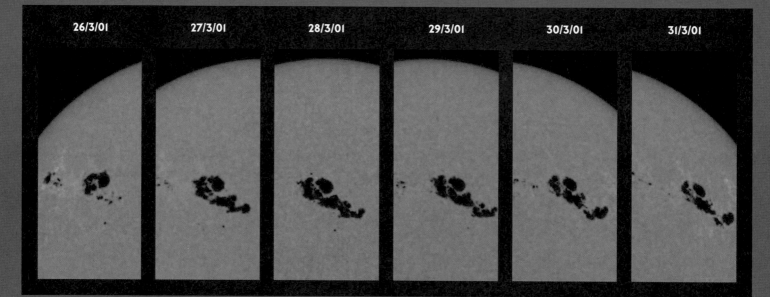

As long ago as 28 BC, Chinese astronomers in the Central Asian deserts had observed dark spots on the surface of the Sun. When the wind blew enough sand into the air to filter the Sun's glare they could see these strange spots and recorded them in the Chinese history book, *The Book of Han*. Over the next 1,500 years many other people recorded these strange dark spots on the surface of the Sun, but it wasn't until the invention of the telescope that Galileo was able to correctly explain the phenomena of sunspots.

The picture on the opposite page was taken by the SOHO probe – the Solar and Helical Observatory that was launched in December 1995. SOHO is giving us unprecedented detail on the life of our Sun and delivering the most beautiful and intricate images of our star that we've ever seen. In the picture above (also taken by the SOHO probe) you can see a beautiful example of the birth, life and death of a sunspot. It may look small compared to the size of the Sun, but the sunspot you are looking at is in fact bigger than the Earth. Sunspots are transient events on the surface of the Sun that are caused by intense magnetic activity that inhibits the flow of heat from deep within the Sun up to the surface. These spots appear dark because they are dramatically cooler than the surrounding area – often 2,000 degrees Celsius cooler. In the eighteenth century it was thought they might even be cool enough to allow humans to land on the surface of a sunspot, but at a toasty 3,000–4,500 degrees Celsius even these cool spots on the Sun would melt a spaceship instantaneously.

Sunspots expand and contract as they move across the surface of the Sun, and they can be as large as

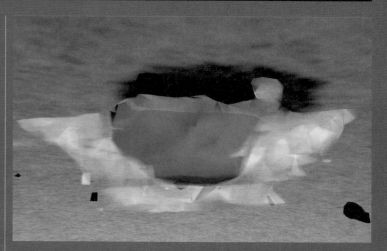

80,000 kilometres (50,000 miles) in diameter, making larger ones visible from Earth without a telescope. Advanced technology and space observation now allows us to track their numbers as they ebb and flow across the face of the Sun.

Since sunspots are cooler areas than on the rest of the Sun, we might have expected to find that the power of the Sun dimmed when the sunspot activity was at its height. In fact, we've found the opposite to be true: the greater the number of sunspots, the more powerful our star becomes. This variation is not just random, as we have studied the Sun in greater and greater detail we have begun to observe patterns emerging; patterns that seem to have direct links to our climate back here on Earth. We've discovered that the Sun has seasons ◉

BELOW AND TOP OPPOSITE:
The dark spots that appear on the surface of the Sun are fleeting blemishes. Sunspots expand and contract as they move across the Sun's surface; these photos were taken using the SOHO probe, but some sunspots are so large that they are visible from Earth without the aid of a telescope.

EARTH

THE SUN AND EARTH: SHARING A RHYTHM?

For decades scientists have sought to understand how these subtle changes in the Sun's power might be affecting the Earth. It's a puzzle that led one man to look away from the Sun and focus instead on the rivers around the Iguaçu Falls. Argentinean astrophysicist Pablo Mauas has spent the last decade analysing data that details every aspect of this river system – from water levels to flow rates – from 1904 all the way through the twentieth century. Unlike many of the world's great rivers, the Parana is so large that it can be navigated by very big ships – and where there are ships there are records. These records enabled Pablo to uncover an extraordinary history and to reveal that, just like sunspots, the river too has a rhythm.

Pablo and his team found that the stream flow of the river fluctuated dramatically three times during the last century, but the records gave no indication as to what was behind these fluctuations. The amount of water in the river Parana seems to follow a pattern, and Pablo had a hunch that this rhythm might be connected to the rhythms of the Sun. To try to link events on the Sun, 150 million kilometres away, to the flow of the great Parana, Pablo looked first at the most obvious rhythm of our star.

We've known for over 150 years (since the German astronomer Heinrich Schwabe collated the data that stemmed back to Galileo's earliest observations of sunspots) that the Sun follows a cycle that is repeated approximately every eleven years. This cycle reflects a rhythmic variation in the number of sunspots, which gives us a very clear indication of the amount of radiation given off by the Sun: the greater the number of sunspots, the greater the energy that is reaching the Earth. But when Pablo Mauas looked for a link between the Paranas rhythm and the eleven-year cycle, at first he found nothing. So instead he turned to calculations that described the Sun's underlying brightness during the last century. We already know that climate change and events such as El Niño can boost the flow of the river, but when Pablo removed both of these effects from the data there appeared to be a strong relationship between the solar data and the stream flow. Superimpose the solar data on the water levels in the river (see below) and you see that when solar activity rises, the volume of water in the river goes up. There is a beautiful correlation between the flow in these rivers and the solar output. Pablo has revealed an amazing link across 150 million kilometres of space that may one day help us to not only better understand the impact of the Sun on our climate, but also to predict the likelihood of floods in the heavily populated waterways of one of South America's greatest rivers.

Changes in the Sun seem to move weather systems elsewhere, too. In India, the monsoon appears to follow a similar pattern to the river Parana, boosting precipitation when solar activity is at its greatest, whereas in the Sahara desert the opposite seems to occur: more solar activity means less rain. The exact mechanisms by which our star may affect Earth's weather remain, for now, a mystery. We know that the energy production rate of the Sun – the power released in the fusion reactions at the core – is very constant, indeed. It doesn't change, as far as we can tell, and so the changes that we see must be to do with the way in which the energy exits the Sun. And while the amount of radiation that falls onto the surface of the Earth is only at the tenths of a per cent level, it really does reveal the intimacy and delicacy of the connection between the Sun and the Earth ◉

WATER LEVELS PLOTTED AGAINST SUN ACTIVITY

WATER LEVELS - - - - -   SUN ACTIVITY - - - - -

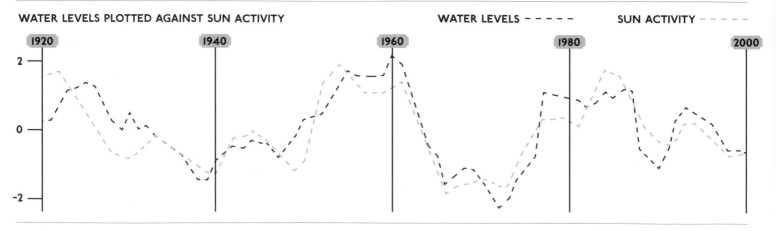

LEFT AND BELOW: Argentinean physicist Pablo Mauas's research suggested that there may be a direct correlation between the activity of the Sun and the flow of the Parana River.

HOW TO CATCH A SUNBEAM

We are tied to our star in the most intimate of ways. All the planets in our solar system are bathed in varying levels of sunlight, but only on one do we know of a phenomenon that does more than just passively receive the warmth of the Sun. Here on Earth, we actually feed on starlight. The Sun is the source of energy for almost all life on Earth; every plant, algae and many species of bacteria rely on the process of photosynthesis to create their own food using the power of the Sun. This in turn creates the foundations for the complex web of life here on Earth; not only does the process of photosynthesis maintain the normal level of oxygen in the atmosphere, but it is also the basis on which almost all life depends for its source of energy.

We are only just beginning to understand the complex mechanisms by which plants capture sunlight; some of this explanation may take us off into the quantum world, but at its most basic chemical level photosynthesis is a simple process. Inside every leaf are millions of organelles called chloroplasts, and it's these little units that do something magical when they capture a photon that has taken the eight-minute, 150-million-kilometre (93,000-million-mile) journey from the Sun. The chloroplasts take in carbon dioxide and water, and by capturing the energy from a sunbeam they convert this into oxygen and complex sugars. It's these complex sugars, or carbohydrates, that are the basis of all the food we eat – whether directly through the consumption of photosynthetic plants or indirectly through the consumption of animals that feed on them. The amount of energy trapped by photosynthesis is immense – around 100 terawatt-years, which is six times larger than the power consumption of human civilisation.

This intimate link between our planet and the Sun is all around us. Yet although we are surrounded by vast swathes of wonderful green machines that are all feeding on the Sun, the leaves and plants that cover so much of the planet do not just rely on any sunlight. In fact, plants are fussy eaters and have evolved to use just a fraction of the sunlight that makes its way through Earth's atmosphere. On the surface of Earth sunlight may appear white,

LEFT AND BELOW: Plants rely on sunlight for survival, and have evolved to gain maximum energy from whatever level of sunlight they receive.

but when you pass it through a prism, you can see that it is made up of all the colours of the rainbow. Different wavelengths of light have different colours – from the blues with the shortest wavelength to the reds with the longest – but it's not just their colour that distinguishes them. The prism reveals the recipe of light that is specific to our star; we see the red, green and blue photons that make up the sunlight all around us and each of these photons has very specific characteristics. The red photons don't carry much energy but there are lots of them, whereas the blue photons, although there are fewer, carry a little more energy. Plants have evolved to gain the maximum energy most efficiently out of the recipe of light our star throws at us, so they don't just use any photons for photosynthesis but only the ones from the red and blue bit of the spectrum.

This intricate relationship between the evolution of plants and our star has had a profound effect on one of the defining features of our planet. When a red or blue photon hits a plant it is absorbed and so those wavelengths of light can no longer bounce back into your eye. Whereas when a green photon hits a plant it is not absorbed but reflected, so this wavelength of light bounces off a leaf and into your eye to create a living world that is defined by one colour more than any other: green. So the verdancy of the forests and jungles that cover our planet is all down to how plants have adapted to the quality of our star's light ◉

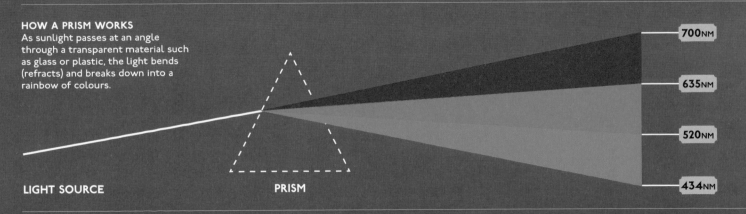

HOW A PRISM WORKS
As sunlight passes at an angle through a transparent material such as glass or plastic, the light bends (refracts) and breaks down into a rainbow of colours.

700NM

635NM

520NM

434NM

LIGHT SOURCE

PRISM

SOLAR ECLIPSE
IN VARANASI

Nothing prepares you for a total solar eclipse, and nothing prepares you for Varanasi. The old Solar City is never quiet and deserted; it is a little slice of ancient India and feels more hectic and vibrant even than the twenty-first-century version. But on the morning of 22 July 2009, the banks of the holy river were packed with people. There was no room, not a square centimetre of space, amongst a million sandaled feet crammed onto the Ghats. Green, yellow, red and orange saris and bronzed torsos bared to the early morning

With immense difficulty, we had found a place to stand in a miraculously under-populated Ghat. We subsequently discovered why it wasn't crowded – it was the public toilet. However, we decided that the unrivalled view of the rising Sun compensated for the smell and we settled down to drink water and wait.

The moment of first contact came at 5.28am, when the limb of the Moon touched the solar disc. The Sun hovered over the river, partially obscured by low cloud, which dimmed the light and made the first moments of the eclipse easier to see. There was little change in the mood of the crowd because, unless you had special cardboard solar sunglasses, there was no perceptible reduction in the Sun's power.

Over the next thirty minutes the Moon's disc quickly slipped across the face of the Sun and I became aware of a strange and unexpected feeling. The Moon moved quickly, and quite unlike the countless other nights I had stared up at its face, it was obviously in orbit – an inhabitant of space rather than a bright disc in the terrestrial sky. I developed a kind of vertigo, because the reality of the Moon as a ball of rock spinning quickly through space transferred to my own situation. I realised that I too was standing on a ball of rock.

By 6.20am, almost an hour after first contact, totality approached. Very, very quickly, the morning light seemed to ebb away, as if time was flowing backwards. But this dimming was not like a sunset because it was so fast. It was not a fading of light; more of a removal. The sound of a million voices dimmed, too, but the Sun still hung as a fainter disc, seemingly unobscured to unshielded eyes. Then at 6.24am, instantly and with Newtonian precision, the Moon slotted into place in front of our star like a perfect Rolexian cog. And quite spontaneously, an enormous cheer erupted from the Ghats.

I then had longer than any TV presenter will have this century to speak to camera about the eclipse. We had worked on words back in London, of course, because we knew this event would be one of the centrepieces of the series, but when the moment came all I could think of was the surprising vertiginous feeling. The red-blue sky of dawn had quickly faded to black as a dark rock swept across a glowing sphere of plasma on its orbital path, leaving me and a million other souls exposed on our own rock to the void. I glimpsed Pascal's terror at the silence of the infinite spaces, turned to camera and said what I felt: 'If you ever needed convincing that we live in a solar system, that we are on a ball of rock orbiting around the Sun with other balls of rock, then look at that. That's the Solar System coming down and grabbing you by the throat.' ◉

summer Sun formed a continuous bridge between the stepped shore and the heavily silted Himalayan waters of the Ganges. The ritualistic instinct to wash in the holy river powered a continuous convective flow of bodies down the concrete steps of the Ghats to the water's edge – a circulating and impenetrable wall of humanity, simultaneously frenzied and calm. As I stood with them I marvelled at the patience of the Indian people – something British film crews dripping with tripods and flight cases will never be able to emulate.

THE INVISIBLE SUN

From 150 million kilometres away the Sun in our sky looks like a perfect disc. It is in fact closer to a near-perfect sphere than any planet or moon in the Solar System; it measures half a million kilometres across, but the variation in its breadth from top to bottom and side to side is little more than ten kilometres. However, this near perfection belies the incredible complexity of the structure. Its constituents are simple enough – to a very good approximation, our Sun is composed of the two simplest elements in the Universe – hydrogen and helium.

RIGHT: Here, the solar eclipse of 1 August 2009 reaches the point of totality as the Moon blocks the Sun entirely. As it does so, the solar corona, invisible at all other times, is clearly on show.

THE STRUCTURE OF THE SUN'S ATMOSPHERE

Hydrogen makes up about three-quarters of the mass of the Sun, with helium making up about a quarter. Less than 2 per cent consists of heavier elements such as iron, oxygen, carbon and neon. This spinning ball of the simplest elements is almost 330,000 times as massive as Earth. It is neither gas, liquid or solid, but a fourth state of matter known as a plasma. A plasma is a gas in which a large proportion of the atoms have had their orbiting electrons removed. This happens because the temperature is high enough to literally strip the atomic nuclei of their electrons. Plasmas are the most common state of matter in the Universe, and in fact we encounter them every day on Earth – fluorescent light bulbs are filled with glowing plasma when they are illuminated. Because plasmas contain a high proportion of naked, positively charged atomic nuclei and free negatively charged electrons, they are electrically conductive and so hugely responsive to magnetic fields.

This gives the Sun a whole host of strange characteristics that are not found on any other body in the Solar System. It rotates faster at its equator than at its poles, with one rotation taking twenty-five days at the equator and over thirty days at the poles.

One hundred and fifty times denser than water and reaching temperatures of up to fifteen million degrees Celsius, the core of the Sun is a baffling and bewildering structure. It is where the Sun's fusion reactions occur, generating 99 per cent of its energy output. Around 600 million tonnes of hydrogen are fused together every second, creating 596 million tonnes of helium. The missing four million tonnes is converted into energy – as much as ninety billion megatons of exploding TNT – and this energy is transported to the surface by the high-energy photons or gamma rays released in the fusion reactions.

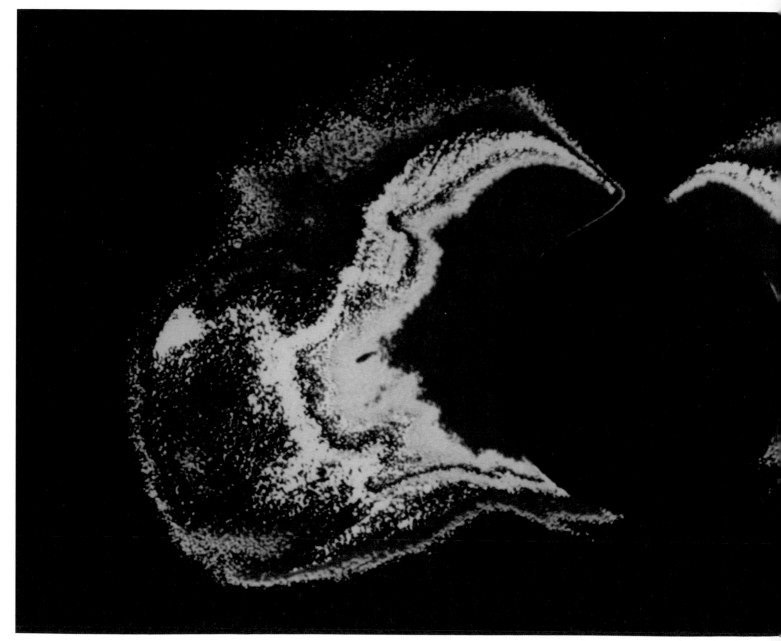

The life of a newly created photon in the core of the Sun is a not a simple one, though. Most are quickly absorbed within a few millimetres of their point of creation by the dense plasma of the core, then they are re-emitted in random directions. In this way the journey of a gamma ray from the core of the Sun to its surface is like a very hot, very long and very unpredictable game of pinball; one that results in the release of millions of lower-energy photons at the Sun's surface. All the light that reaches us here on Earth is incredibly ancient; it is estimated that a single photon can take anywhere from 10,000 to 170,000 years to make the journey from the Sun's core to surface before it can make the eight-minute journey into our eyes.

By the time a photon reaches the surface, or photosphere, the Sun's temperature has dropped from thirteen million degrees Celsius to about 6,000 degrees. It's this massive change in temperature that causes the vast convection currents that swirl through the Sun, creating thermal columns that carry hot material to the surface and create its characteristic granular apeparance we see from Earth.

This is only just the beginning of the story of our sun's mighty physical presence. Beyond the surface of thc Sun is the strange and invisible layer known as the solar atmosphere. Only visible to the naked eye on Earth during a total solar eclipse, the Sun's atmosphere is made up of a thin collection of electrically charged particles, protons and electrons. Unsurprisingly, the atmosphere of the Sun cools as you get further away from the surface. At a distance of 500 kilometres (310 miles) is an area known as the Temperature Minimum, which has a temperature of around 4,400 degrees Celsius. This location, as the name implies, is the coolest area of our star and the first place in which we can find simple molecules like water and carbon dioxide surviving in close proximity to the Sun. Beyond this region something odd happens. As you move further away from the Sun into space, the atmosphere doesn't get cooler, it gets dramatically hotter. This outer region of the Sun's atmosphere is known as the corona. This mysterious layer of the Sun only becomes visible to the naked eye during a total solar eclipse but when it is revealed you

One hundred and fifty times denser then water and reaching temperatures of over 13 million degrees Celsius, the core of the Sun is a baffling and bewildering structure.

are seeing a structure that is larger and hotter then the Sun itself. With an average temperature of a million degrees Celsius and some areas reaching colossal temperatures of up to twenty million degrees Celsius, this vast cloud is, in places, hotter than the core of the Sun. The mechanisms that drive the corona to these high temperatures are not yet fully understood, but this effect is certainly due to the complex magnetic interactions that occur between the surface and the corona. What is known is that each and every day, at the very top of the atmosphere, some of the most energetic coronal particles are escaping. The Sun leaks nearly seven billion tonnes of corona every hour into space; a vast superheated supersonic collection of smashed atoms that en masse are known as the solar wind. This is the beginning of an epic journey that will see the Sun's breath reach the furthest parts of the Solar System, creating the final vast structure of our star – the heliosphere ◉

LEFT: The solar corona is the outer atmosphere of the Sun, which stretches out more than one million kilometres from its surface. It is visible to us on Earth as a halo around the Sun, but only during a total solar eclipse when the Sun's surface is obscured.

THE HELIOSPHERE

The heliosphere is a gigantic magnetic bubble in space that contains our solar system, the solar wind and the entire solar magnetic field. This bubble extends far into the Solar System, possibly even forty to fifty times further from the Sun than the Earth, and is shaped by the solar winds coming from the Sun.

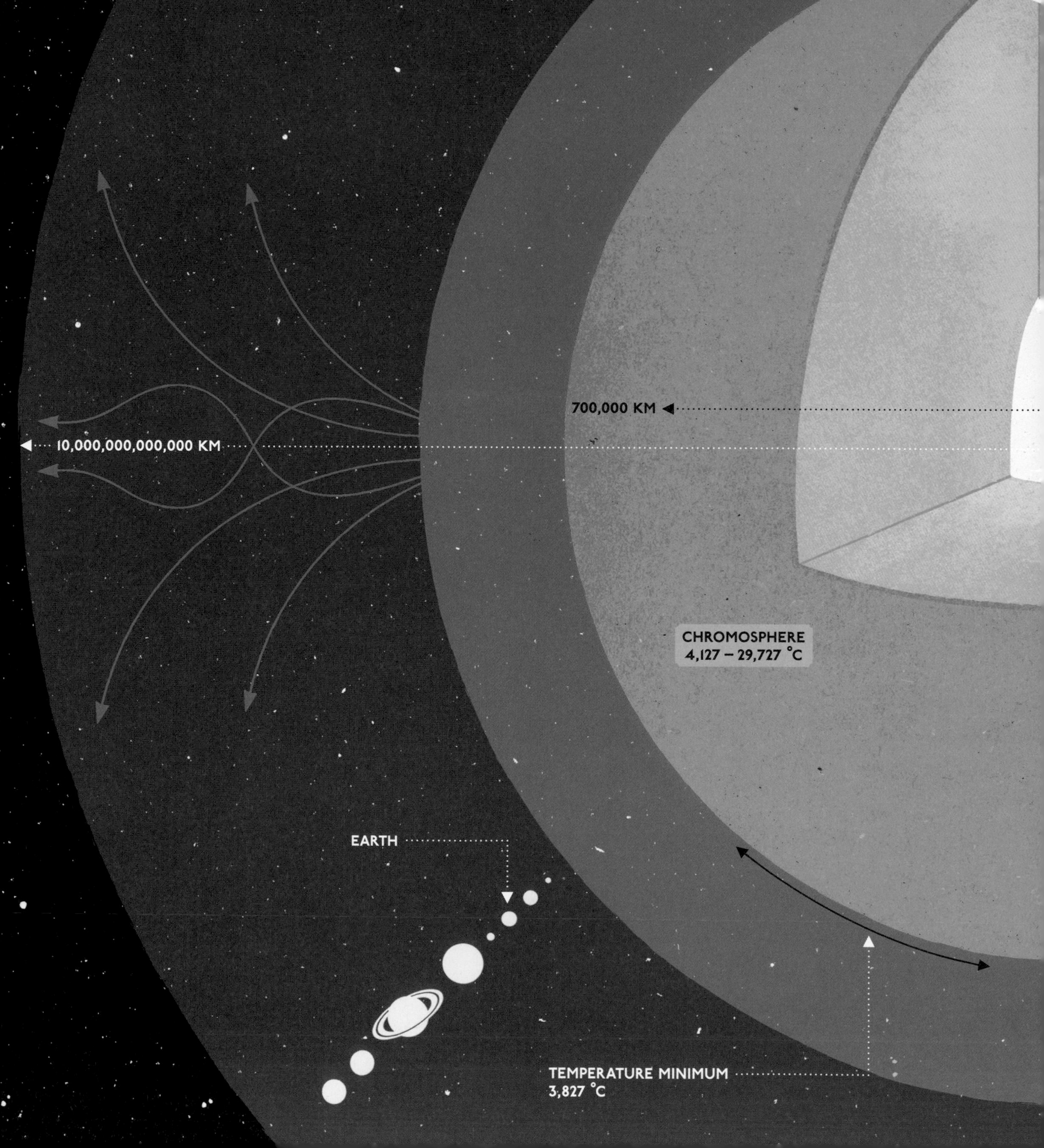

700,000 KM ◄

10,000,000,000,000 KM

CHROMOSPHERE
4,127 – 29,727 °C

EARTH

TEMPERATURE MINIMUM
3,827 °C

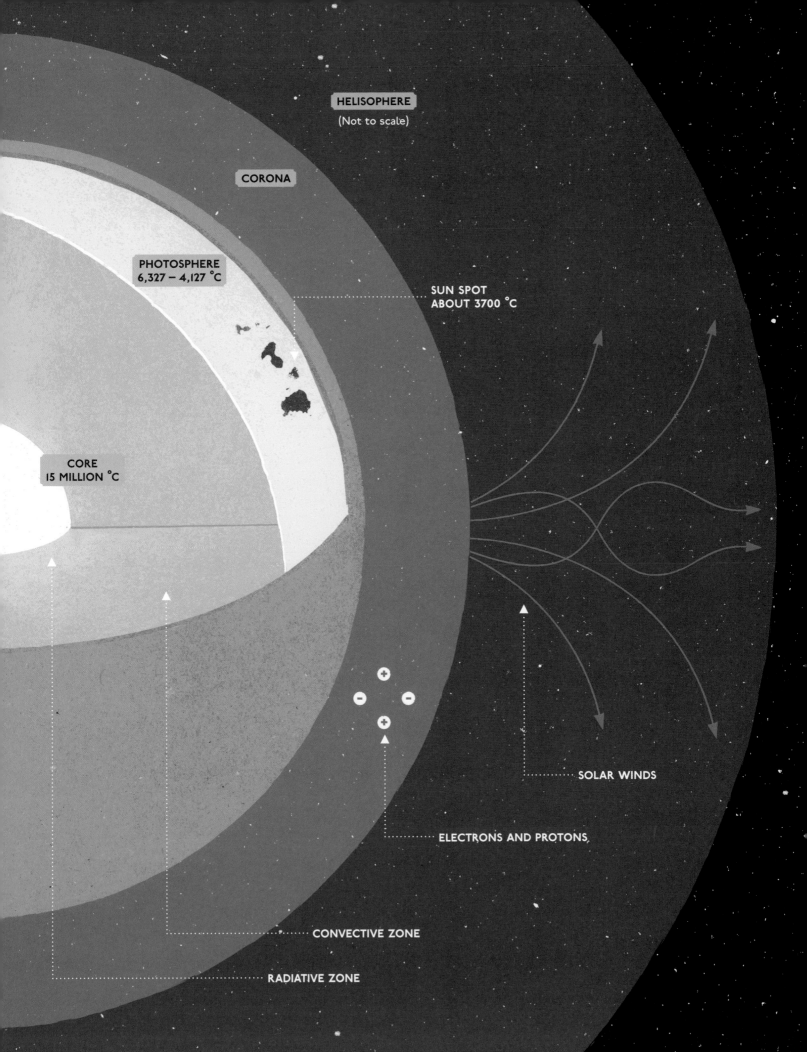

HELISOPHERE
(Not to scale)

CORONA

PHOTOSPHERE
6,327 – 4,127 °C

SUN SPOT
ABOUT 3700 °C

CORE
15 MILLION °C

SOLAR WINDS

ELECTRONS AND PROTONS,

CONVECTIVE ZONE

RADIATIVE ZONE

RIGHT: This image of a solar flare was captured by the TRACE spacecraft on 21 April 2002. Solar flares are explosive bursts of gas that release huge amounts of energy. In the geomagnetic storm that follows each flare, particles and radiation are hurled out into space.

On a beautiful sunny winter's day in the Arctic, it's hard to imagine that our star could be a threat. Yet, high above us deadly solar particles stream our way at speeds topping a million kilometres an hour and bombard the Earth.

DEFENCE AGAINST THE FORCE OF THE SUN

Astronomy has a long history of discoveries by amateurs. From Clyde Tombaugh, the man who discovered Pluto, to David Levy, the co-discoverer of the Shoemaker-Levy comet, the freedom of the skies has always tempted non-professionals to bypass the experts and break new ground. Amateur British astronomer Richard C. Carrington is a worthy member of this list. In 1858, Carrington made the first observation of an event that would eventually become known as a solar flare.

This massive explosion in the Sun's atmosphere releases a huge amount of energy and Carrington noticed that this event was followed by a geomagnetic storm, a massive disruption in the Earth's magnetic field, the day after the eruption. Carrington was the first to suspect the two events may be connected. Beyond the weather in our swirling atmosphere, the solar wind creates another more tenuous atmosphere and weather system that surrounds our planet. We rarely notice this ethereal weather high above us, because by the time the solar wind reaches Earth it's pretty diluted. If you went into space close to the Earth and held up your hand, you wouldn't feel a thing. In fact, there are about five protons and five electrons for every sugar cube's worth of space, but they're travelling very fast, carrying a lot of energy – enough to severely damage our planet's atmosphere, were it not for a defence system generated deep within the Earth's core.

On a beautiful sunny winter's day in the Arctic, it's hard to imagine that our star could be a threat. Yet, high above us deadly solar particles stream our way at speeds topping a million kilometres an hour and bombard the Earth. Down here on the surface we're protected from that intense solar wind by a natural shield that deflects most of it around us. To see that shield, you need nothing more than a compass. That's because the Earth's force field is magnetic, an invisible shell that surrounds the planet in a protective cocoon.

The magnetic field emanates from deep within our planet's spinning iron-rich core. It's this gigantic force field, known as the magnetosphere, that deflects most of the lethal solar wind harmlessly away into space. However, the planet doesn't escape completely; when the solar wind hits the Earth's magnetic field, it distorts it. It stretches the field out on the night side of the planet and in some ways it's like stretching a piece of elastic. More and more energy goes into the field and over time this energy builds up, stretching the tail until it can no longer hold on to it all. Eventually the energy is released, accelerating a stream of electrically charged particles down the magnetic field lines towards the poles. When these particles, energized by the solar wind, hit the Earth's atmosphere, they create one of the most beautiful sights in nature: the aurora borealis, or Northern Lights ◉

LIGHT FANTASTIC –
THE AURORA BOREALIS

The northern Norwegian city of Tromso is known as the gateway to the Arctic. At latitude seventy degrees North, deep inside the Arctic Circle, it has permanent sunlight from mid-May until the end of July, and permanent darkness from late November to mid-January. In late March the Arctic Ocean is a dark frosty blue, the white-crested waves matching the layers of snow and ice packed solid onto the wooden jetties and the well-weathered decks of the fishing boats. It was an utterly magical place to begin filming on 22 March 2009.

We had gone to see the aurora borealis. Tromso is perfectly positioned on the auroral arc – the thin circle around the North Pole along which the elusive light show usually appears. March and September are the best months to see it, due to the alignment of the Earth's magnetic field relative to the Sun, and we were told that, given clear skies, we would have a strong chance of glimpsing the Northern Lights.

Our guide told me of a Sami legend about the aurora. (The Sami are the people of the North, whose domain stretches from Tromso in the west, across northern Sweden

The Northern Lights reveal in exquisite beauty our planet's connection with the rest of the Solar System. The Earth's environment does not end at the edge of our atmosphere; it stretches at least to the Sun.

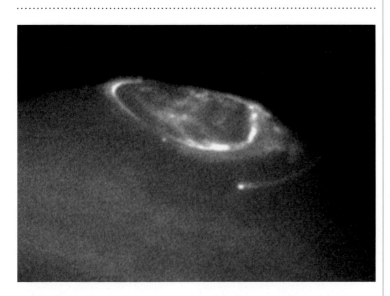

LEFT: Waiting for the Northern Lights in Norway – we travelled well out of the city in the hope of witnessing this fantastic natural light show.

LEFT, TOP: The Northern Lights were a spectacular sight that were well worth the wait. The crystal-clear skies were punctuated by green shafts of light that seemed to rise up from the mountain range.

ABOVE: Jupiter has the largest and most powerful magnetic field in the Solar System, so aurorae are a permanent fixture around the planet's poles and have also been seen around its moons.

and Finland and into Russia.) The legend has it that the aurorae are the spirits of women who died before they had children. Trapped between the frozen land and heaven, they are condemned to dance forever in the dark Arctic skies. As dusk fell, we rode snowmobiles out into the dense forests by the Fjord to get away from the city lights and settled down in the Sami camp with hot reindeer stew to wait.

Just after midnight, the aurora came. I walked out into the frigid night air, enjoying the crunch of footsteps in fresh snow, and looked up. They came gently, a vague hint of green, but built quickly; sheets of colour drifted slowly then suddenly broke off and danced impossibly fast, a three-dimensional rain of light rising and falling between land and sky. They were mostly green, with hints of orange and red close to the horizon. They were like nothing I have ever seen, and as I turned to camera I realised that I didn't care about the physics of what I was seeing. My reaction, composed whilst sitting at my desk in Manchester, was worthless in the face of Nature at its most magnificent. The Sami had it right – an aurora isn't the light shaken out of atoms of nitrogen and oxygen as they are bombarded by high-energy particles from Earth's ionosphere accelerated down magnetic field lines towards the poles, it is made of majestic, mournful, dancing spirits, trapped in the Arctic night.

The Northern Lights reveal in exquisite beauty our planet's connection with the rest of the Solar System. The Earth's environment does not end at the edge of our atmosphere; it stretches at least to the Sun. We are bound to our star by the visible light that creates and nurtures life on Earth and the unseen, constant solar wind that only appears to us at night in special circumstances. Each and every planet in the Solar System shares this connection, and the same laws of physics apply. As the solar wind races out into the Solar System, wherever it meets a planet with a magnetosphere aurorae spring up. Jupiter's magnetic field is the largest and most powerful in the Solar System, and the Hubble space telescope reveals that there are permanent aurorae over the Jovian poles. Jupiter's moons, Io, Europa and Ganymede also have aurorae, created by Jupiter's atmospheric wind interacting with the moons' atmospheres. Saturn too puts on an impressive display, with aurorae at both its poles, but because Saturn's magnetic field is uneven, the aurorae are smaller and more intense in the north.

As the solar wind reaches the edge of the heliosphere it begins to run out of steam. Incredibly, there is a probe out there that will discover where these solar winds end ◉

BELOW: Aurora Borealis, natural light displays which are only visible from the northern hemisphere, are on show here over Lake Thorisvatn in Iceland.

VOYAGERS' GRAND TOUR

BELOW: Voyager 1 has been of vital importance to space exploration for over thirty years. The craft was responsible for this picture of the Moon and the Earth, which was the first of its kind taken by a spacecraft, recorded on 18 September 1977.

RIGHT: On 5 September 1977, Voyager 1 was launched from Kennedy Space Center at Cape Canaveral, Florida. Although its twin spacecraft, Voyager 2, was launched sixteen days earlier, Voyager 1 followed a faster flight path, which meant it reached Jupiter first.

In the autumn of 1977, a pair of identical 722-kilogramme (1,592-pound) spacecraft were launched from Cape Canaveral, Florida. Voyagers 1 and 2 were about to embark on a very special mission: to visit all four of the Solar System's gas giants – Jupiter, Saturn, Uranus and Neptune. Normally such a journey would take thirty years to complete, but by a stroke of good fortune these spacecraft were designed at a time when the planets were uniquely aligned, allowing the probes to complete their grand tour in less than twelve years. Today, over thirty years after their launch, both spacecraft are alive and well, and remarkably Voyager 1 is still reporting back to Earth – the ultimate and most wonderful example of mission creep in the history of space exploration.

Voyager 1 is currently the furthest man-made object from Earth. Travelling at seventeen kilometres (eleven miles) per second, this extraordinary spacecraft is just over seventeen billion kilometres (eleven billion miles) from home and delivering knowledge that it was never designed or expected to uncover. Listening to Voyager 1 is the sensitive ear of the Goldstone Mars station in the Mojave desert, California; one of the few telescopes in the world that is capable of communicating over such vast distances. Voyager is so far away that it takes the signal around fifteen hours to arrive, travelling at the speed of light. It may appear as little more than a blip on a screen, but the information Voyager is sending is providing the first data from the frontier of our solar system, from the edge of the heliosphere, and constantly measuring the solar wind as it fades away. Voyager 1 has now reached the point where this wind that emanated so powerfully from the surface of the Sun has literally run out of steam. The heliopause is the boundary at which the solar wind is no longer strong enough to push against the stellar winds of the surrounding stars. Beyond this point Voyager will leave its home and head off into interstellar space. With the batteries expected to struggle on until 2025, this spacecraft will continue to feed us data as it becomes the first man-made object to leave our solar system ◉

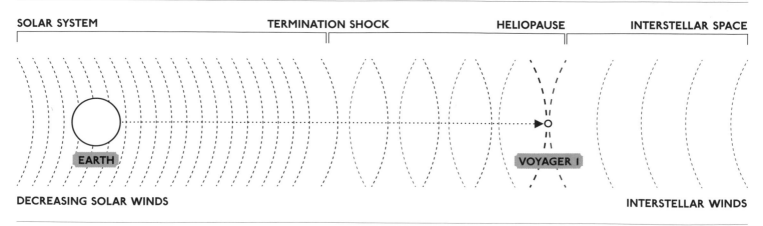

| SOLAR SYSTEM | TERMINATION SHOCK | HELIOPAUSE | INTERSTELLAR SPACE |

EARTH

VOYAGER 1

DECREASING SOLAR WINDS

INTERSTELLAR WINDS

FROM EARTH TO THE OORT CLOUD

BELOW: You could be forgiven for missing the small red star in the middle of this image. Its light is so faint that this dwarf star, Proxima Centauri, the nearest star to the Sun, was only discovered in 1915.

BOTTOM: The Oort cloud is an almost spherical collection of icy objects that is believed to lie approximately one light year away from the Sun. Gravitational pull from other stars can cause these icy objects to enter the Solar System as comets.

Our journey through the Sun's Empire doesn't end at this distant frontier, seventeen billion kilometres (eleven billion miles) away, where the solar wind meets the interstellar wind. The Sun has a final, invisible force that reaches out much further. Our star is by far the largest wonder in the Solar System. In fact, it alone makes up 99 per cent of the Solar System's mass. It is this immensity that gives the Sun its furthest reaching influence – gravity.

This is the full extent of the Sun's empire; the lightest gravitational touch that retains a cloud of ice that encloses the Sun in a colossal sphere. Beyond this Oort cloud there is nothing. Only sunlight escapes; light that will take four years to reach even the Sun's closest neighbour, Proxima Centauri – a red dwarf star among the 200 billion others that make up the Milky Way. And it's by looking here, deep into our local galactic neighbourhood, that we're learning to read the story of our own star's ultimate fate ◉

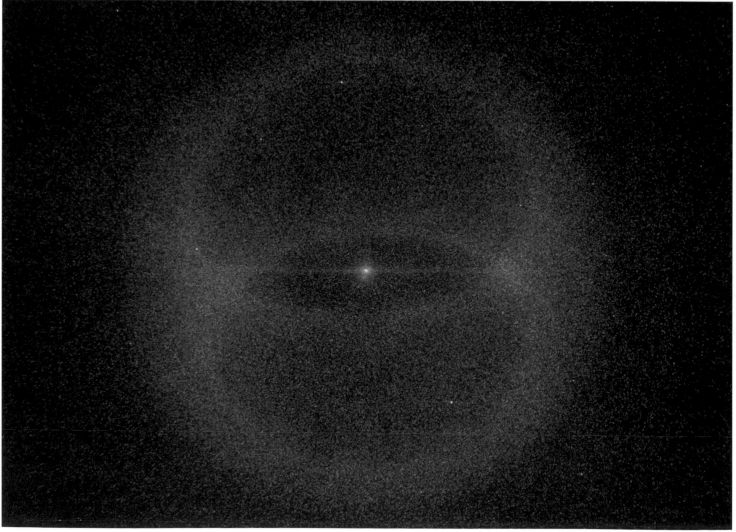

INVESTIGATING THE FUTURE OF OUR SUN

The Sun's empire is so vast and so ancient, and its power so immense, that it seems an audacious thought to imagine that we could even begin to comprehend its end – the death of our sun. However, that is exactly what astronomers are trying to do, and many of them head to the most arid and barren desert on Earth, the Atacama, in Chile, looking for answers.

There, high up on an the sides of an extinct volcano at an altitude of 2,635 metres (8,643 feet), sits Paranal Observatory, home to the world's most powerful array of telescopes. On arrival we were given 'important information for a safe stay on Paranal'. As the observatory is about two and a half kilometres (one and a half miles) in the air, we were advised that if we experienced any of the following, we should consult a paramedic immediately: headache and dizziness, breathing problems, ringing or blocking of the ears, or seeing stars. It honestly said that if you saw stars at the Paranal Observatory you should consult a paramedic immediately!

Perched high above the clouds is the reason why so many astronomers venture to this desert. Here, four colossal instruments make up the European Southern Observatory's 'Very Large Telescope', or VLT. If you look up at the sky with these mighty machines you quickly notice that the stars are not just white points of light against the blackness of the sky, but are actually coloured. Through these lenses, orangey-red, yellow and bluey-white stars fill the clear Chilean sky.

However, this beauty is not just one of the wonders of our night sky, it has also revealed something much deeper. To gaze upon the galaxy full of stars is to observe them at all the stages of their lives – from youthful bright stars to middle-aged yellow stars very similar to the Sun. Contained within the night sky we can see a colour code that allows us to plot the life cycle of every star, including our own ◉

If you look up at the sky with these mighty machines you quickly notice that the stars are not just white points of light against the blackness of the sky, but are actually coloured. Through these lenses, orangey-red, yellow and bluey-white stars fill the clear Chilean sky.

ABOVE LEFT: The European Southern Observatory's Very Large Telescope consists of four 8-metre (26-foot) aperture telescopes that work independently or combined as one of the world's most powerful telescopes.

ABOVE: The Very Large Telescope is famous for its high level of observation and its spectroscopic resolution. This spectacular image of the Orion Nebula demonstrates the exceptional work of this machine.

THE HERTZSPRUNG-RUSSELL DIAGRAM

For the last 100 years astronomers have meticulously charted the nearest ten thousand stars to Earth and arranged each according to its colour and brightness. From this was born the Hertzsprung-Russell diagram; a powerful and elegant tool that allows astronomers to predict the history and evolution of stars, and in particular the future life of our sun. Most of the stars, including our own, are found in the 'main sequence' – the band of stars that runs from the top left to the bottom right. The Sun will spend most of its life there, steadily burning its vast reserves of hydrogen fuel, which will last for another five billion years. After which, it will pass through a Red Giant phase.

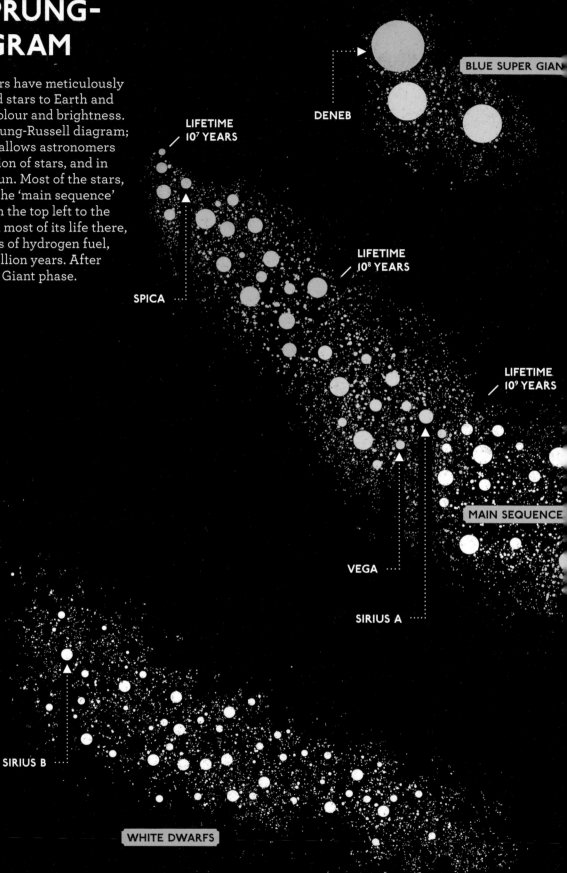

BLUE SUPER GIANT

DENEB

LIFETIME
10^7 YEARS

LIFETIME
10^8 YEARS

LIFETIME
10^9 YEARS

SPICA

MAIN SEQUENCE

VEGA

SIRIUS A

SIRIUS B

WHITE DWARFS

←— INCREASING   25,000 °C   10,000 °C

TEMPERATURE °C

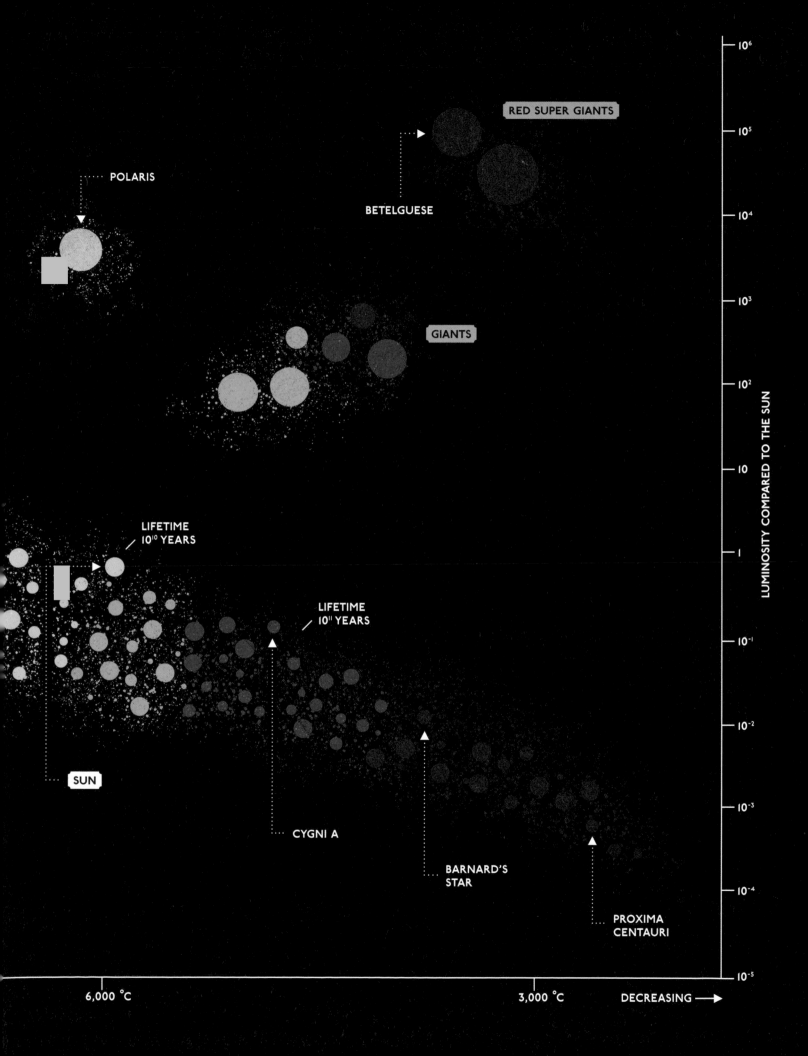

EMPIRE OF THE SUN

THE DEATH OF THE SUN

Eventually, like all stars, the Sun's fuel will run out, its core will collapse and our star will begin its final journey. At this stage you may expect it to slowly burn out and splutter its way into oblivion, but there is a final, remarkable twist to our Sun's ten-billion-year story.

When the fuel does finally run out, the nuclear fusion reactions in the Sun's core will grind to a halt and gravity will be master of our star's fate once more. The Sun will no longer be able to support its own weight and it will begin to collapse. Just as in its formation, this collapse will start to heat the Sun once more, until the layers of plasma outside the core become hot enough for fusion to begin again – but this time on a much bigger scale. Our star's brightness will increase by a factor of a thousand or more, causing it to swell to many times its current size. The Sun will then drift off the main sequence and into the top right-hand side of the Hertzsprung-Russell diagram, into the area known as the Giant Branch.

As the outer layers expand, the temperature of the surface will fall and its colour will shift towards red. Mercury will be little more than a memory as it is engulfed by the expanding red sun, which will grow to two hundred times its size today. As it swells the Sun will stretch all the way out to the Earth's orbit, where our own planet's prospects are dim.

So it seems that the wonder that has remained so constant throughout all of its ten billion years of life will end its days as a giant red star. For a few brief instants the Sun will be two thousand times as bright as it is now, but that won't last long. Eventually our star will shed its outer layers and all that will be left will be its cooling core – a faint cinder, or White Dwarf, that will glow pretty much to the end of time, fading slowly into the interstellar night. As it does so, all its wonders – the aurorae that danced through the atmospheres of planets of the Solar System, and its light that sustains all the life here on Earth – will be gone.

The gas and dust of the dying Sun will drift off into space, and in time they will form a vast dark cloud primed and full of possibilities. Then, one day, another star will be born, perhaps with a similar story to tell, the greatest story of the cosmos ◉

CHAPTER 3

ORDER OUT OF CHAOS

THE CLOCKWORK SOLAR SYSTEM

The story of the Solar System is the story of the emergence of order out of chaos, guided by the simplest law of physics: gravity. The planets and their moons exist in relatively stable orbits because of a delicate interplay between gravity and angular momentum, and this beautiful natural balance is written before our eyes in the spinning patterns and rhythms of the heavens.

In the small ancient city of Kairouan on the north-eastern plains of Tunisia lies the fourth most holy site in the Islamic world. Founded by Arabs in 670 CE, this city of just 150,000 people is home to the oldest place of Muslim worship in the Western world. The great mosque of Kairouan is both impressive in its beauty and also in its scale. Covering over 9,000 square metres (97,000 square feet), the Mosque resembles a great fortress as well as a place of worship. At its heart is a vast courtyard and near the centre is a beautiful piece of astronomical engineering – an ancient sundial. Humans have used sundials like this one to follow the brightest star in the heavens for over 5,500 years.

For the last fourteen centuries, the sundial at the centre of this great mosque has measured the relentless passing of the days, marking out the passage of time as the Sun travels across the sky, plotting the call to prayers before dawn, at sunrise, at noon, at sunset and in the evening.

The sundial is a beautifully simple piece of technology. Originally nothing more complicated than a stick in the ground or the length of a human shadow, sundials have enabled us to measure time by following the movement of the Sun across our sky. For thousands of years this movement appeared to confirm the Earth's position at the centre of the Universe. From the most simple of observations it seemed to make perfect sense that the Sun orbits the Earth every 23 hours and 56 minutes. Yet the simple, regular rhythm that each one of us witnesses every day is nothing more than an illusion. It is not the Sun that's moving, what we are observing is the rotation of the Earth as it travels through space.

BELOW: The great mosque at Kairouan is the oldest in Africa, and at the heart of its spectacular central courtyard is a beautiful piece of astronomical engineering – an ancient sundial.

It's wonderful to think that across the planet the rhythms of our lives are governed by our journey through space. From waking up to going to bed, eating strawberries in July or a tangerine in December.

BELOW: The powerful effect of the Solar System on our climate is seen at this french market. The stall is laden with seasonal strawberries and mushrooms.

Travelling at 108,000 kilometres an hour, on a 900-million-kilometre journey around the Sun, our planet completes this epic journey once every 365.25 days. The year in the life of our planet is just one of the endless rhythms by which we live our life – and all of these are governed by the seemingly clockwork motion of our planet. It carries us through cycles of night and day as it turns on its axis, rotating at 1,700 kilometres an hour every twenty-four hours. The length of the day at a particular place on the Earth's surface is dictated by the precise angle of our planet in relation to the Sun.

We have seasons here, too, due to the fact that the Earth's axis is tilted by twenty-three degrees. As we journey around the Sun this angle creates the changing dynamic that defines the cycles of many of the creatures that live both on the land and in the oceans. In the Northern Hemisphere the summer months coincide with the North Pole leaning towards the Sun; when, at this time of year, the angle favours the northern half of our planet with extra energy from our star. By winter the dynamic has changed; the North Pole is pointing away from the Sun and the Southern Hemisphere is bathed in additional sunlight.

It's wonderful to think that across the planet the rhythms of our lives are governed by our journey through space. From waking up to going to bed, wearing a jumper one month and a T-shirt the next, eating strawberries in July or a tangerine in December, each of these everyday events is intimately connected to a journey through space that catapults us at 108,000 kilometres (67,000 miles) per hour around a star, but leaves most of us completely unaware of this rollercoaster ride through the cosmos.

It's not only Earth that is subject to these rhythms – the whole Solar System is full of these cycles, with each planet orbiting the Sun at its own distinct tempo. Mercury is the fastest; closest to the Sun, it reaches speeds of 200,000 kilometres (124,000 miles) per hour, completing its orbit in just eighty-eight days. Venus rotates so slowly that it takes longer to spin on its axis (225 days) than it does to go around the Sun, so that on Venus (and also on Mercury) a day is longer than a year. Further out, the planets orbit more and more slowly. Mars completes one orbit of the Sun every 687 days, just a couple of months short of two Earth years. Jupiter, the largest planet, takes twelve Earth years to complete each orbit, Saturn almost thirty years, Uranus eighty-four years and at the very furthest reaches of the Solar System, four and a half billion kilometres from the Sun, Neptune travels so slowly that by 2009 it hadn't completed a single orbit since it was discovered in 1846.

The Solar System is driven by these rhythms, so regular that the whole thing could be run by clockwork. It seems extraordinary that such a well-ordered system could have come into being spontaneously, but this little island of order that we call our solar system is in fact a wonderful example of the beauty and symmetry that emerges as a result of the action of the simple physical laws that govern the Universe. Studying these laws has lead us not only to understand how that order emerged from the chaos of space, but has also helped us to understand the origins and formation of the Solar System itself. Understanding our position in the Solar System is one of the great journeys of science and is a story that is as old as human civilisation ◉

RHYTHMS OF THE SOLAR SYSTEM

All the planets in the Solar System are subject to specific rhythms and cycles and they are all as regular as clockwork. Each planet orbits the Sun at its own distinct tempo; however, the closer the planet is to the Sun, the faster it completes its orbit.

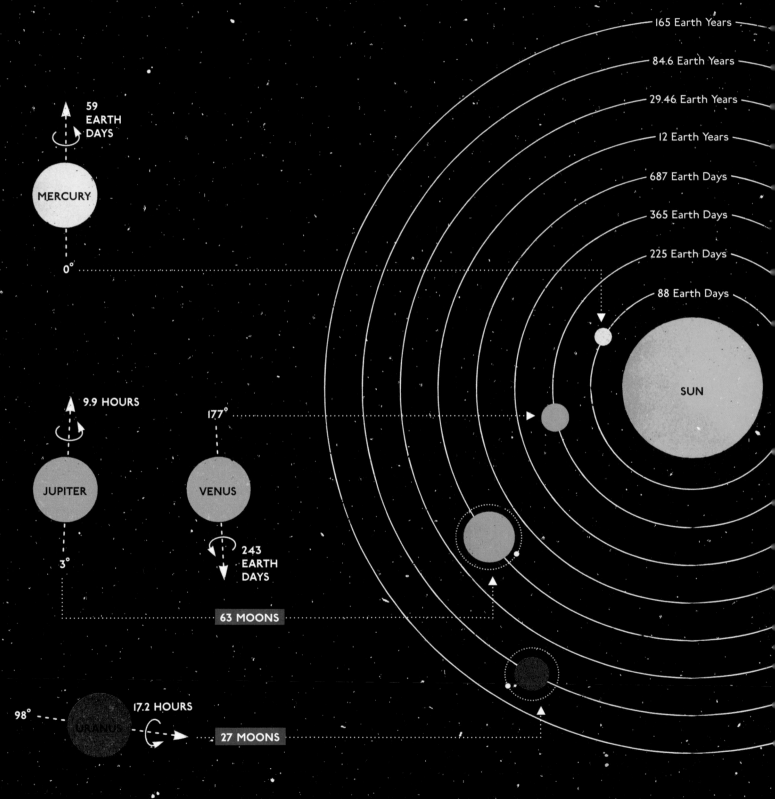

59 EARTH DAYS

MERCURY

0°

9.9 HOURS

JUPITER

3°

177°

VENUS

243 EARTH DAYS

63 MOONS

98°

URANUS

17.2 HOURS

27 MOONS

165 Earth Years

84.6 Earth Years

29.46 Earth Years

12 Earth Years

687 Earth Days

365 Earth Days

225 Earth Days

88 Earth Days

SUN

UNKNOWN SEASONS

MILD VARIATION BETWEEN SEASONS

EXTREME SEASONS

SEASONS SIMILAR TO OURS

EARTH

24.6 HOURS

16 HOURS

MARS

NEPTUNE

2 MOONS .... 25°

28°

13 MOONS ....

10.7 HOURS

SATURN

62 MOONS ....

27°

24 HOURS

Northern
Spring /
Southern
Autumn

Northern
Winter /
Southern
Summer

Northern
Summer /
Southern
Winter

Northern
Autumn /
Southern
Spring

1 MOON .... 23.5°

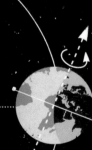

THE CENTRE OF
THE UNIVERSE

According to Greek mythology, Atlas was the powerful God who carried the Earth on his shoulders, supporting the heavens from the Atlas mountains in North Africa. Today these mountains are still one of the finest places to view the stars. City life may have robbed us of our connection to the night sky, but from the inky black of the Atlas mountains it's easy to appreciate the profound effect it would have had on our ancestors. They looked into the sky to understand their place in creation, and the movement of the stars told them one thing: they were at the centre of the Universe.

Watch the night sky for any length of time and it's no surprise they came to this conclusion. The North Star, Polaris, almost exactly aligned with the Earth's spin axis, adds to the illusion that all the stars are rotating through the sky at that point. It's what the ancients thought for thousands of years, but even though it appears obvious, it is, of course, wrong.

To understand the Earth's real position in the Solar System we need to look at the one set of bodies that doesn't

ABOVE: Star trails in the night sky over Tunisia. The movement of the stars traces spectacular concentric arcs.

RIGHT: In this combination of images taken over a few months, we can trace the movement of Mars in the Earth's sky. Generally it travels in a straight line, but once every two years Mars is passed by Earth and the planet appears to be moving backwards – a phenomenon known as retrograde motion.

behave as predictably as the stars. The Greeks named them planets, or wandering stars, and we have kept the name planet to describe them.

The picture of Mars opposite shows how, rather than travelling in a straight line across the background of the stars, the planet occasionally changes direction and loops back on itself. The Greeks had observed this strange movement as long ago as 1534 BCE but could not explain it. Why would this planet create such strange patterns in the sky if it was orbiting around the Earth at the centre of the Universe? Explaining this retrograde motion of Mars didn't come easy; in fact, it took over 3,000 years to reach the right answer.

For half that time the work of one man, Claudius Ptolemaeus, dominated our view of the Universe. In around 150 CE Ptolemaeus published his great work, *The Almagest* – a complete explanation of the complex movement of the planets and stars. For well over a thousand years the Ptolemic view of the Solar System was set in stone, with the Earth at the centre and everything else revolving around it. This was science and religion working hand in hand. With man as the most important of God's creations, it was only right that the Earth should be at the centre of a perfect and uniform universe. Something seemingly as inexplicable as the retrograde motion of Mars was explained by the presence of smaller orbits known as epicycles, which perfectly matched the observed data.

This tendency, perhaps innate in humans, to accept the word of authority in earthly and heavenly matters, has been one of the great obstacles to progress throughout history. The Royal Society in London, the oldest scientific society in the world, was formed in 1660 and took on the motto '*Nullius in verba*', which translates as 'Take nobody's word for it'. In other words, a true and deep understanding of the Universe is gained not by reading the authoritative words of the ancients, but by careful observation of Nature and original thinking.

To drag the Earth away from the centre of the Solar System was the life's work of a Polish astronomer, Nicolaus Copernicus, one of the founding fathers of modern science. Copernicus quite literally turned our view of the Solar System inside out. Using the most basic of instruments, he collected the data that would lead him to create an entirely new view of the heavens, the heliocentric view – one that placed the Sun very firmly at its centre ◉

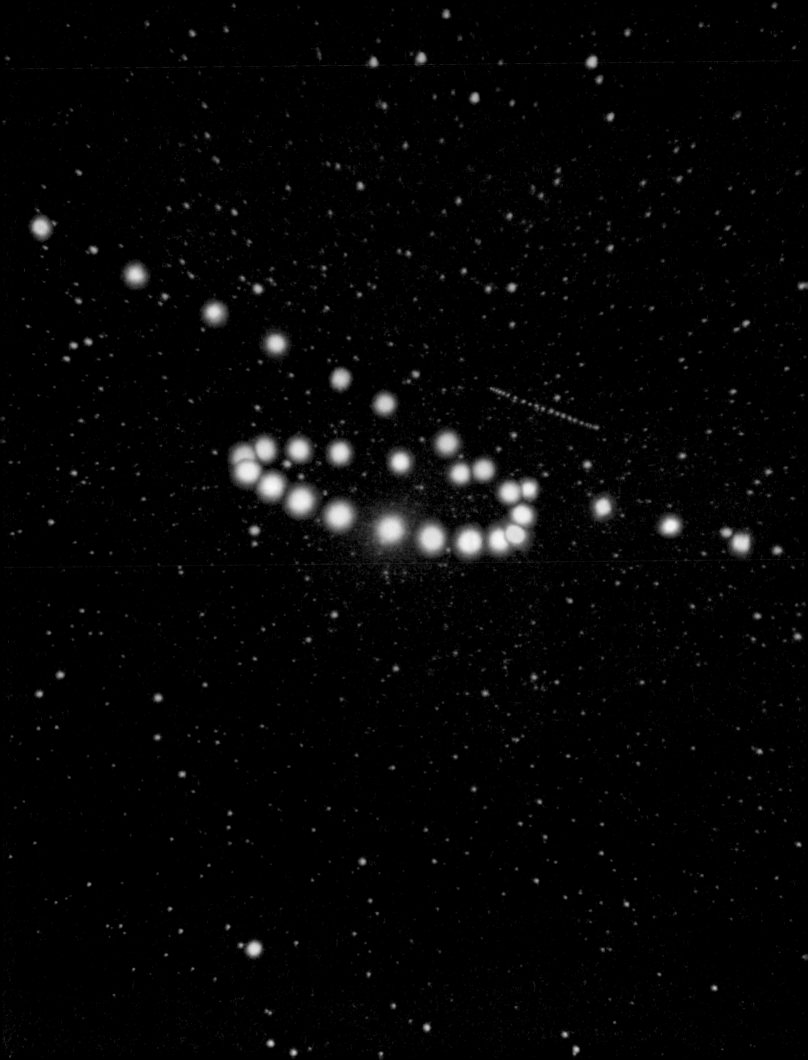

THE HELIOCENTRIC VIEW

Although completed over a decade earlier, Copernicus' theory was only published shortly before his death in 1543. *On the Revolutions of the Celestial Spheres* took apart 1,500 years of astronomical thought, replacing it with a new way of thinking. At its heart was an explanation for the mysterious retrograde motion of a planet like Mars.

The diagram opposite explains his theory. With the Sun at the centre of the Solar System and the planets going around it in almost circular orbits, you suddenly find we are sitting in a very different seat here on Earth to watch the celestial clockwork. As we speed around the Sun together, Mars appears to move in a straight line across the sky. It makes sense that as Mars and Earth move we appear as two speeding cars next to each other on a motorway. Remember that Mars is travelling slower than Earth, twenty-four kilometres (fifteen miles) per second to Earth's thirty kilometres (nineteen miles) per second, and so Earth catches then overtakes Mars and suddenly our perspective is shifted. Earth then pulls away from Mars, which appears to move backwards against the fixed stars – it seems to reverse its direction. Just like a car overtaking on the inside, the slower planet appears to move in reverse.

Mars continues to appear to move backwards across the sky until Earth has sped so far in front that the perspective changes once again and Mars regains its usual direction of travel across the night sky. So Mars seems to have executed a strange looping motion in the sky because Earth overtook it on the inside, which is why we see retrograde motion. It's simple when explained, but it took millennia to work out!

Understanding the retrograde loops was one of the major achievements of early astronomy. It created the concept of the Solar System and allowed us to build the first accurate maps of the planets and their orbits around the Sun. Once we had this picture, many new questions arose. Building on the work of Copernicus, generations of scientists have explored the fundamental workings of our solar system. In doing so they were forced to ask profound questions about its origins, such as: why is it so ordered, and how did that order emerge from the chaos of the heavens?

To find answers we must search for clues in what we see. A good place to start is the circular motions of the planets. The explanation for this astronomical clockwork lies beyond our solar system, because it requires an understanding of the physical principles that govern the whole Universe. To understand the origins of our solar system, we need to look around us ◉

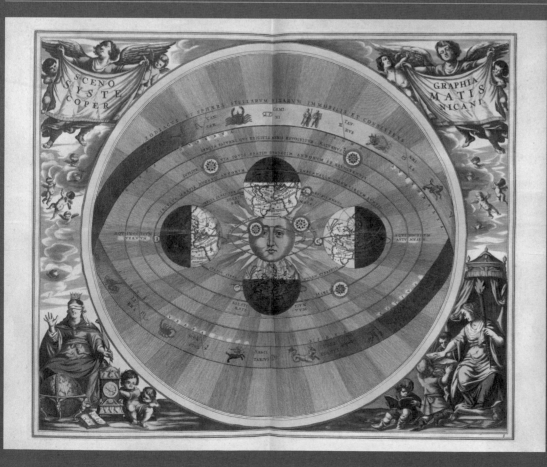

LEFT: This engraving by Andreas Cellarius in 1660 illustrates the theory of Nicolaus Copernicus' world system. This essentially suggests that the Sun is at the centre of the Universe, around which the other stars move in a circular, uniform motion.

MARS IN RETROGRADE
When in retrograde, Mars moves backwards with respect to the fixed star positions behind it. It seems to travel in a loop as Earth overtakes it from the inside.

EARTH   MARS

FIXED BACKGROUND OF STARS

10   9   8   7   3   4 2 6 5   1

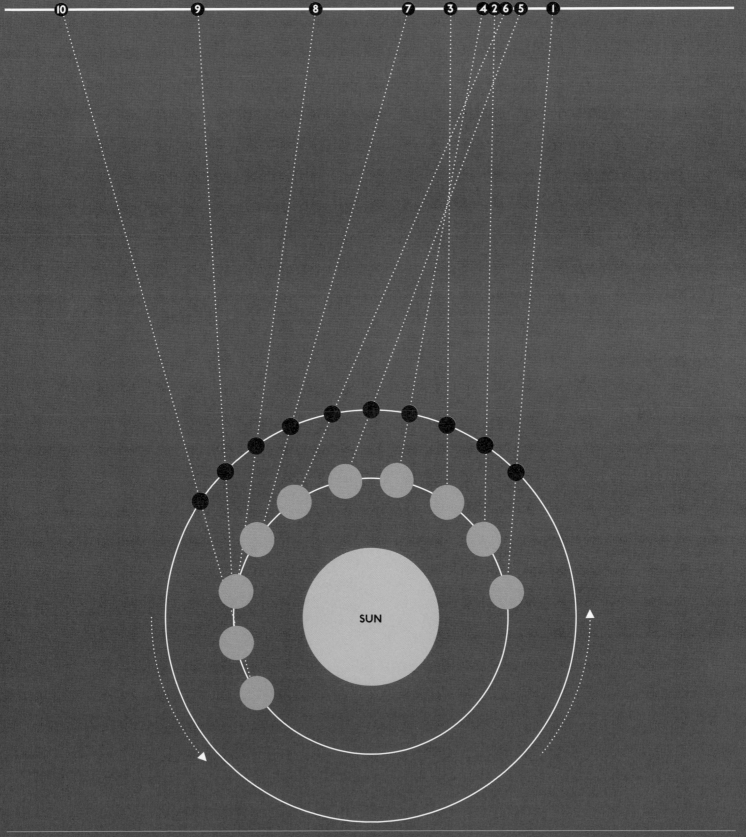

SUN

THE BIRTH OF THE SOLAR SYSTEM

BELOW AND BOTTOM: Lying in the middle of what is known as Tornado Alley, the US state of Oklahoma experiences hundreds of tornadoes between April and June every year. These giant rotating storms tear across the ground, leaving chaos and destruction in their wake.

One of the remarkable things about the laws of Nature is that they are universal. In other words, the same laws that describe the formation of the Solar System must also describe the most mundane things on Earth. The majestic spinning motions of the planets as they journey around the Sun must therefore be described by the same laws as other things that spin – like the seemingly ordinary motion of water as it spirals down out of a sink. Spinning spirals are seen all over the Earth and all across the Universe. We see them everywhere because the laws of physics are the same everywhere.

Each year in Oklahoma these universal laws unleash forces that drive some of the most powerful and destructive phenomena on our planet. Oklahoma lies in the middle of a part of the United States known as Tornado Alley, where between April and July hundreds of twisters tear across the landscape. They are incredibly dangerous and destructive, and their key feature is a violent, spinning column of air.

For professional storm chasers, the challenge is to get as close to a tornado as possible. However, playing with the most intense of all atmospheric phenomenon does come loaded with risk. A tornado can pick up a car and throw it half a kilometre through the air, crushing it into a ball. This immense destructive power is generated by intense low pressure, which seeds the high wind speeds and rapid rotation that characterise a twister.

Tornadoes are most often born from a type of thunderstorm known as a supercell. These giant rotating storms start high in the atmosphere, but as they develop they descend towards the ground, sucking up hot air and contracting into a tightly spinning funnel. The wind speed increases as the storm contracts, in order to obey one of the most fundamental of universal principles, known as the conservation of angular momentum. On Earth, if the conditions are right, the conservation of angular momentum in these colossal collapsing storm systems can rapidly generate wind speeds of at least 300, sometimes 400, and in exceptional cases 500, kilometres an hour.

ANGULAR MOMENTUM

To a physicist, quantities that are conserved are of overwhelming importance. A conserved quantity is something that never changes – something that can be neither created nor destroyed. Energy is an example of such a conserved quantity, but there are others. One with which we are probably all familiar, although it may have an unfamiliar name, is linear momentum.

Momentum to a physicist is the speed of something (or, more accurately, its velocity, which is its speed in a particular direction) multiplied by its mass – 'mv'. Imagine a cannon sat waiting to be fired. Both the cannon and the cannon ball are still. This means the momentum of both is zero because nothing has a speed. Now fire the cannon. The cannon ball flies out at high speed and therefore has a momentum equal to its velocity times its mass. But momentum is a conserved quantity, which means it cannot be created or destroyed. This means that even as the cannon ball is flying through the air, the combined momentum of the cannon and the ball must still be zero. It is, because the cannon recoils in the opposite direction to the ball, and the sum of its and the ball's momentum will be zero. The cannon doesn't fly backwards at the same speed as the ball because its mass is bigger, and it is only the product of its mass and speed that must balance – bigger mass, smaller speed, same momentum. You might ask what happens to all the momentum when the cannon ball hits the ground or the cannon stops recoiling. It isn't destroyed – the ball transfers its momentum to the Earth as it ploughs into the ground, and transfers a bit to molecules of air that it hits as it flies through the air. The cannon will transfer its momentum to the molecules in the ground through friction. If we were sufficiently clever we could track the movements of all the molecules jiggled around by the cannon and ball, and the momentum of everything combined would always add up to zero.

Linear momentum has a counterpart called angular momentum. Instead of measuring the speed with which something flies in a straight line, angular momentum deals with the speed at which something spins. For mathematicians: if something of mass 'm' is flying in a circle at a distance 'r' from the centre with a (tangential) velocity 'v', the angular momentum about the centre of the circle is 'mvr' (see below).

Like linear momentum, angular momentum is conserved – it too can be neither created nor destroyed. A classic example of the conservation of angular momentum in action is a spinning ice-skater. If the skater starts to spin then pulls her arms inwards, she will spin faster. This is because the angular momentum of her hands is bigger the further they are from the centre of her spin – in this case, her body. If she pulls her arms in, therefore, angular momentum from her hands is lost, and this must be compensated for by an increase in the spin rate of the rest of her body – she speeds up.

Again, the observant reader may have noticed a subtlety here. How did the skater start to spin if angular momentum can't be created? The answer is that the skater pushed against the ice to start spinning, and the ice is connected firmly to the Earth. Just as the cannon recoiled against the cannon ball to conserve angular momentum, the entire Earth recoils against the spinning skater to conserve angular momentum and the Earth's spin rate changes. This is a miniscule effect, of course, because the Earth is many millions of times bigger than the skater. As the skater slows down through friction with the ice, just as for the cannon, the angular momentum is redistributed back to the Earth by friction. The total spin of everything never changes, though; angular momentum is always conserved.

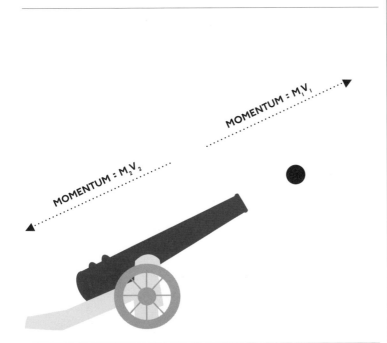

ONE PRINCIPLE OF GENERAL RELATIVITY
(i) Linear momentum is conserved. (ii) Momentum before =0. Therefore (iii) m1v1 + m2v2 =0. (iv) Heavy cannon recoils slowly against fast moving cannon ball.

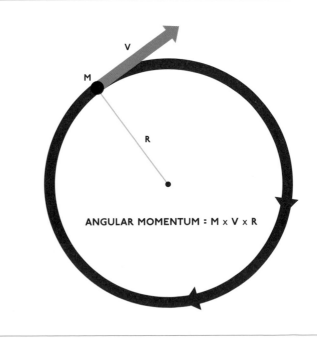

CONSERVATION OF ANGULAR MOMENTUM
If the mass (M) is constant, any decrease in the length (R) results in an increase in the velocity (V) – hence why a skater spins faster when she pulls her arms in, effectively reducing her R.

BELOW: In 1995, NASA's Hubble space telescope captured the image of three space pillars, dubbed 'The Pillars of Creation'. These dramatic pillars in the Eagle Nebula were formed over millions of years from radiation and dust from around twenty or more huge stars.

OPPOSITE: The dark spots in this picture reveal a type of interstellar cloud known as a Bok globule. Bok globules are the least understood object in the Universe, despite extensive studies by the astronomer Bart Bok, from whom they get their name. These small, cold, gas and dust clouds condense to form stars.

Bizarre as it sounds, the universal processes that shape these vast storm systems are the same as those that sculpted the early Solar System, because the laws of physics are universal and apply equally to everything. The Eagle Nebula, an interstellar cloud of dust, hygrogen and helium and a sprinkling of heavier elements, is about 6,500 light years away from Earth. It is almost 100 trillion kilometres (60 trillion miles) tall and within its towering pillars, stars are made. This famous photograph (below), known as 'The Pillars of Creation', was taken by the Hubble space telescope in 1995. As well as being stunningly beautiful, it also gives us a glimpse of where we came from. Five billion years ago everything we know and see around us was formed from a nebula just like

this – a giant cloud of gas and dust. Drifting across light years of space, that cloud remained unchanged for millions of years, until something happened that caused it to coalesce into the Solar System we have today. It is thought that a supernova, the explosive death of a nearby star, sent shockwaves through the nebula and caused a clump to form in the heart of the cloud.

Ths clump would have been denser than the surrounding cloud, so its gravitational pull would have been stronger. Very slowly, over millions of years, it would have pulled in more and more gas and dust. Eventually the whole cloud would have collapsed in on itself faster and faster, but crucially, as it collapsed it was doing something that would establish a series of events that we all experience today: the cloud was spinning.

is one of the coldest known objects in the natural universe. Its low temperature is critically important, because it means that the gas and dust molecules are moving very slowly and are therefore more easily captured by the weak force of gravity, and pulled together. As the collapse continued, one particularly dense area dominated and the gas slowly formed a more familiar shape. Out of the cold a new star was born.

While gravity caused one part of our early solar system to contract further and further, until nuclear fusion reactions halted the collapse, the rest of the spinning disc was stabilised by a different mechanism. It was the conserved spin that balanced the inward pull of gravity. Imagine you throw a tennis ball through the air. It forms an arc in a curved path

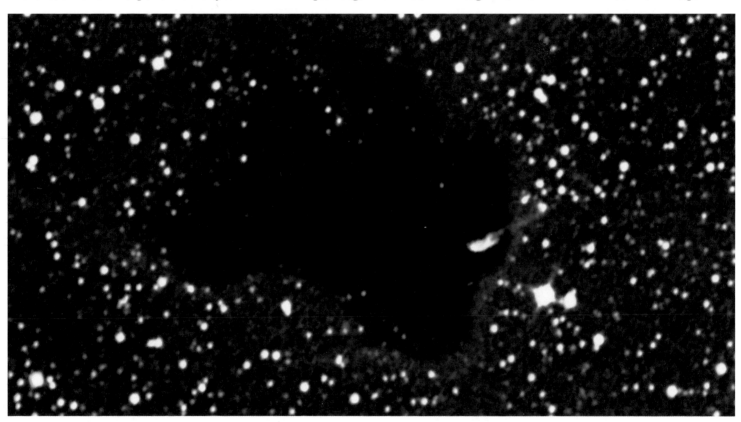

The initial spin would probably have come from the glancing blow dealt by the supernova shockwave. Just like the spinning vortex of a tornado, this spinning cloud of cosmic dust from which all of us emerged had to follow the universal laws of the cosmos. It doesn't matter if you are a tiny molecule of air on a minuscule planet, or a vast cloud of gas containing all the ingredients of a solar system, if something spinning contracts then the universal principle of conservation of angular momentum dictates that it must rotate faster.

In the case of a tornado this contraction unleashes incredibly destructive forces. The core rotates faster and faster and a column of violently rotating air descends from the cloud, wreaking havoc on any man-made structures that it encounters. Strangely and wonderfully, the universal principle responsible for this violence is also responsible for creating the stability of the Solar System, because it is angular momentum that stops the Solar System collapsing completely.

As the cloud collapsed, one of the strangest and least understood objects in the Universe was formed. A Bok globule

called a parabola, upwards at first, but then back downwards to the ground, where it lands some distance away from you. Unless you throw it vertically upwards, the tennis ball has angular momentum relative to the Earth. This angular momentum would be conserved were it not for the fact that the ball bumps into the ground. But imagine that you could throw the ball very fast – so fast that by the time it moved downwards to the ground the Earth had curved away beneath it. The tennis ball would then be in orbit around the Earth, and if there were no air resistance, no forces would act upon it other than gravity because it would never bump into the ground. It would be constantly falling towards the Earth, and be constantly missing! Because angular momentum, or spin, is conserved, this is a completely stable situation – gravity ensures that the ball keeps falling, and as long as no other forces act, the ball keeps missing the ground and stays in orbit. In exactly the same way, a planet doesn't fall into the Sun even though it is only feeling the attractive force of gravity; it is constantly falling towards the Sun but constantly missing ◉

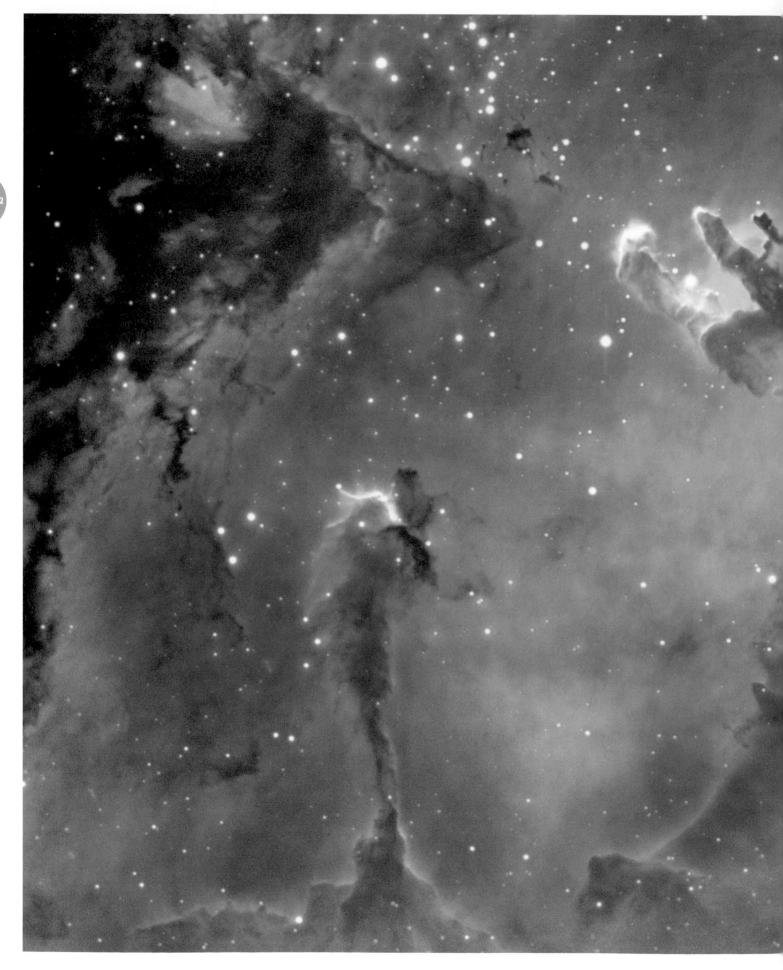

This is how our Solar System was born: rather than the whole system collapsing into the Sun, a disc of dust and gas extending billions of kilometres into space formed around the newly shining star. In just a few hundred million years, pieces of the cloud collapsed to form planets and moons, and so a star system, our Solar System, was formed. The journey from chaos into order had begun.

LEFT: This image of the Eagle Nebula – so-called because it looks like an eagle from afar – gives us a remarkable insight into the birth of a solar system. The tall pillars and round balls of dust and gas signify where new stars are being formed.

WONDERS OF THE SOLAR SYSTEM

SATURN:
THE INFANT SOLAR SYSTEM

Of all the Solar System's wonders, there is a place we can go to where the processes that built the Solar System are still in action today. It is a place of outstanding beauty and complexity; a place that has entranced astronomers for centuries. It is the planet Saturn.

SATURN

1.4 billion kilometres away from the Sun, Saturn sits
six planets away from the centre of the Solar System. It
takes almost thirty years to orbit the Sun, and a day on
Saturn is only around ten and a half hours long, although
it is believed to fluctuate more than our stable days on
Earth. Saturn is the second largest planet after Jupiter,
but while it dwarfs the Earth in volume, it is only ninety-
five times the mass of our home planet. This is because
of its surprisingly low density. One of the more amusing
facts about our solar system is that Saturn would float in
water, if you could find an ocean big enough. Saturn's low
density is a result of its composition; it is mainly composed
of hydrogen and helium, with small amounts of other
trace elements. Along with the other three outer planets –
Jupiter, Uranus and Neptune – Saturn is a gas giant.

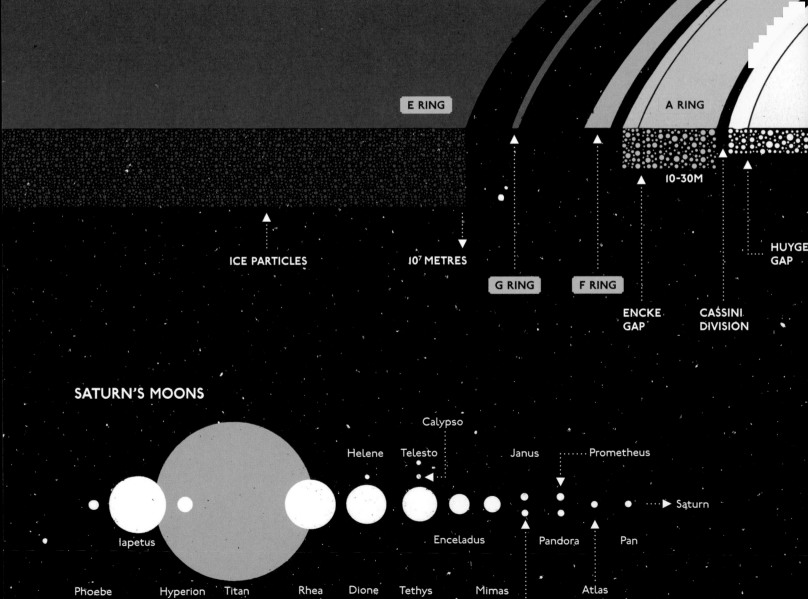

E RING

A RING

10-30M

ICE PARTICLES

10^7 METRES

HUYGE
GAP

G RING   F RING

ENCKE
GAP

CASSINI
DIVISION

SATURN'S MOONS

Calypso

Helene   Telesto   Janus   Prometheus

Saturn

Iapetus

Enceladus   Pandora   Pan

Phoebe   Hyperion   Titan   Rhea   Dione   Tethys   Mimas   Atlas

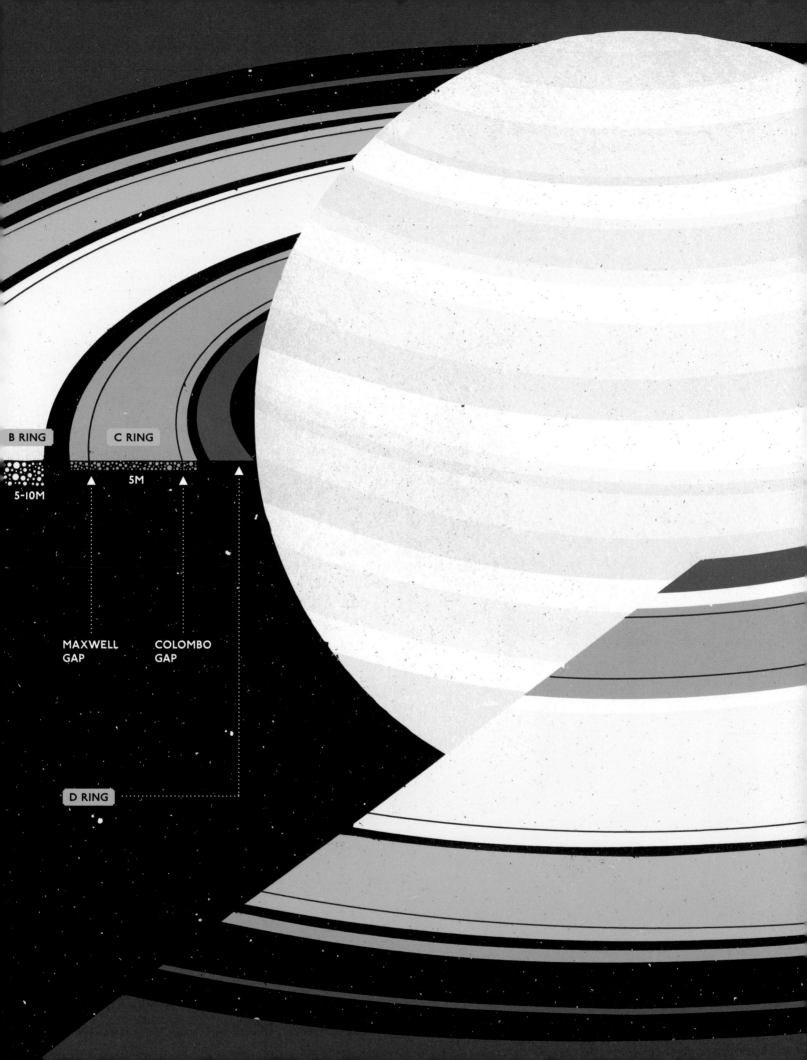

B RING

C RING

5M

5-10M

MAXWELL
GAP

COLOMBO
GAP

D RING

SATURN'S RINGS

Humans have known about Saturn since they first stared up at the sky, because it appears as a bright and beautiful, wandering yellow star. 1.3 billion kilometres (8 billion miles) from Earth, it is the furthest planet visible with the naked eye and has had a special place in ancient mythology for millennia. For all that time and for all the eyes that were trained on this wandering yellow star, though, not a single person was aware of the one characteristic that today draws us to Saturn more than to any other planet. Not until Galileo first trained his telescope on Saturn in 1610 did humans see one of the true wonders of the Solar System: Saturn's rings.

Galileo thought the rings were two moons nestling on either side of the giant planet, and it was forty-five years before the Dutch astronomer Christopher Huygens, using a more powerful telescope that magnified the planet fifty times, distinguished 'a thin flat ring' for the first time. Huygens was also the first to discover Saturn's moon, Titan, but it was the great Italian astronomer Giovanni Cassini who unravelled the first details of the rings' intimate structure. In 1675, having already discovered four more of Saturn's smaller moons – Iapetus, Rhea, Tethys and Dione – Cassini discovered a gap in the rings, now known as the Cassini division.

For the last 350 years, Huygens and Cassini have had their names intimately linked with Saturn. In 1997, this legacy came to a fitting climax with the launch of one of the most extraordinary and ambitious missions ever sent into the outer solar system. The launch of the robotic spacecraft mission, Cassini-Huygens, was controlled from NASA's laboratory in Pasadena, California.

On 1 July 2004, after a 3.5-billion-kilometre (2-billion-mile) journey that included slingshots around Venus and flybys past Earth and Jupiter, Cassini became our one, and to date only, spacecraft in orbit around Saturn. Its purpose is to study Saturn and its rings in such detail that Giovanni Cassini could only have dreamt of. It may take eighty minutes for one of the Cassini spacecraft's images to be sent by radio waves across space to us here on Earth, but again and again the wait has been worth it. For over six years the spacecraft has been sending back the most amazing pictures. They have revealed that the rings are impossibly intricate, made up of thousands upon thousands of individual bands and gaps and surrounded by a network of moons. Part of Cassini's mission is to discover how the rings came to be like this; how all this incredible structure was created. This is interesting in itself, of course, but there is a deeper reason for studying Saturn and her rings: their intricate structure is as

LEFT AND BELOW: Saturn is well known for its distinctive appearance. However, it took early astronomers years to identify Saturn's rings, believing them to be moons. The picture opposite was taken by the Cassini spacecraft, launched in 1997; this and the Cassini division were named after Giovanni Cassini, who discovered the rings in 1675.

The more we can understand the forces at work in the rings, the more we can piece together the origins of our own existence.

close as we can get to the disc of dust, rock and ice that surrounds the primordial Sun, and out of which the planets formed. That is why the Saturnian system has so much to teach us.

Professor Carl Murray is part of the Cassini mission's imaging team and has spent a lifetime studying Saturn's rings. For him the rings are far more than just one of the most beautiful sights in our cosmic backyard, they are a unique opportunity to understand the origins of our solar system through direct observation. 'They are like a miniature solar system because the moons are the equivalent of the planets and Saturn is the equivalent of the Sun', he told me when we met and stood in the gallery above the control room of Cassini. 'The physical processes that go on in the rings and their interaction with the small moons that are around them are probably similar to what went on in the early solar system after the planets formed.' In Murray's eyes, looking at the rings of Saturn is like looking at the Solar System four and a half billion years ago, with the Sun at the centre surrounded by a disc of dust not unlike Saturn's rings.

It is this similarity, the history of our solar system contained within the rings, that convinces Murray of the importance of studying them. 'If we can't understand a disc of material that's in our own backyard, what chance do we have of understanding a disc that's long since disappeared?'

The more we can understand the forces at work in the rings, the more we can piece together the origins of our own existence – and that means understanding the structure of the rings in intricate detail. Now, for the first time, by using the data from Cassini we are able to recreate almost every aspect of the rings. We can journey from the vast scale of the disc to the minute structure of individual ringlets. We can measure the immense speeds of the individual rings as they orbit Saturn. Like the planets orbiting the Sun, the rings nearest Saturn are the fastest, travelling at over eighty thousand kilometres an hour. While the rings appear solid, casting shadows onto the planet, they are also incredibly delicate; the main disc of the rings is over a hundred thousand kilometres across, and less than one kilometre thick.

Saturn's rings are undoubtedly beautiful, and when you see the magnificent pictures sent back from Cassini, it's almost impossible to imagine that that level of intricacy, beauty and symmetry could have emerged spontaneously, but emerge spontaneously it did. For that reason alone, Saturn's rings are a wonder of the Solar System, but there is more to it than that, because in studying the origin and evolution of Saturn's rings, we have at last begun to gain valuable and unprecedented insights into the origins and evolutions of our own solar system ◉

SATURN'S RINGS
ON EARTH

There were two things the boat driver told me about these icebergs when we visited Iceland: one was that they can come up from the bottom of the seabed without any warning, fly up to the surface, tip the boat over and then you die. The other was that if you collect some ice and take it home, it's absolutely brilliant in whisky because the water is pure, a thousand years old, with no pollutants in it and it makes whisky taste superb. So it's either death or whisky. That's my kind of pond!

It's difficult to imagine the scale, beauty and intricacy of Saturn's rings here on Earth, but the glacial lagoons in Iceland can transport our minds across millions of kilometres of space and help us to understand the true nature of the rings. There, as the ancient glacier tumbles off the mountain, huge chunks of ice break off and fill the lagoon below with icebergs as far as the eye can see. At first sight, the lagoon appears to be a solid sheet of pristine ice, but that is an illusion. The surface is constantly shifting, an almost organic, ever-changing raft of thousands of individual icebergs floating on the water. The structure of Saturn's rings is similar, because despite appearances the rings aren't solid. Each ring is made up of hundreds of ringlets and each ringlet is made up of billions of separate pieces. Captured by Saturn's gravity, the ring particles independently orbit the planet in an impossibly thin layer.

But the similarity doesn't end with the layout. The reason Saturn's rings are so incredibly bright from Earth is because they are predominantly made of glacially pure water ice, sparkling as they reflect the faint sunlight; billions of pieces of frozen water, a billion kilometres away from Earth. Most of the pieces are smaller than a centimetre, many are micron-size ice crystals, but some are as big as an iceberg, some are as big as houses and some can be over a kilometre across.

What a wonderful thing it would be to stand on one of the larger pieces of Saturn's rings and gaze out over thousands of kilometres of glistening ice. The lagoon is perhaps as close as a human being will get for hundreds of years, but its beauty allows the imagination to fly. I hope that, one day, we will follow.

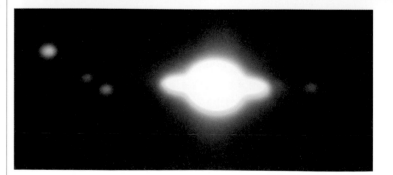

TOP: The structure of Saturn's rings is remarkably similar to the way these beautiful, shifting icebergs float in this Icelandic lagoon.

ABOVE: A truly incredible sight; Saturn and its rings viewed through a telescope – you can even make out several moons around the planet.

YOUNG OR OLD

The brightness of the ice in Saturn's rings is a puzzle because our solar system is a very dirty place. It is full of dust, and until recently it was thought that Saturn's rings must be relatively newly formed, because otherwise their magnificence would have been dimmed as the surfaces of the ice crystals became coated in this interplanetary debris. It is now thought that they are much older – hundreds of millions or even billions of years old. The reason the rings shine so brightly is that, like the icebergs in the lagoon, the rings are constantly changing. Paradoxically, despite their intricate, almost eternally beautiful structure, we now know that the rings are a chaotic place, at least in the small scale of individual ice particles. As the ring particles orbit Saturn, they crash into each other and collect into giant clusters that are endlessly forming and breaking apart. As they collide, the particles shatter, exposing bright new faces of ice that catch the sunlight. It is because of this constant recycling that the rings are able to stay as bright and shiny as they were when they formed.

This dynamism is one of the most remarkable and surprising things about Saturn's rings. Their constant renewal is why they are clean enough to reflect sunlight and allow us to see them at all. If they were static, their magnificence would have long ago been dimmed by dust. They are different today than they were a thousand years ago; they'll be different in a hundred or a thousand years' time, too, but that structure and that beauty, that magnificence, will remain, possibly for as long as there is a solar system ◉

Each ring is made up of hundreds of ringlets and each ringlet is made up of billions of separate pieces. Caught within the grasp of Saturn's gravity, the ring particles independently orbit around the planet in an impossibly thin layer.

The reason Saturn's rings are so incredibly bright from Earth is because they are made of almost pure water ice, sparkling in the sunlight, billions of these pieces, a billion kilometres away from Earth.

As the ring particles orbit Saturn, they're continually crashing into each other and collecting into giant clusters that are endlessly forming and breaking apart. As they collide, the particles shatter exposing bright new faces of ice that catch the sunlight. It's because of this constant recycling that the rings are able to stay as bright and shiny as they were when they formed.

Most of the pieces are smaller than a centimetre, many are micron-size ice crystals, but some are as big as an iceberg, some are as big as houses and some can be over a kilometre across.

THE MOONS OF SATURN

When Galileo first looked up at Saturn through his primitive telescope in 1610, he thought the planet had ears. With a telescope no more powerful than a pair of modern binoculars, Galileo could only guess at what he was seeing. Having recently discovered four moons orbiting Jupiter he came to the obvious conclusion – Saturn had two giant moons orbiting the planet and the 'lobes' he could see were simply an illusion. To his dismay, as the celestial dance between Earth and Saturn progressed under his watchful eye, these 'moons' seemed to disappear; Saturn, it seemed, had 'swallowed' its children. What Galileo was actually seeing were the rings of Saturn disappearing as Saturn and Earth danced around the Sun, shifting our perspective so that the rings were edge-on to us.

Although his first observation was wrong, Galileo's guess that Saturn, like Jupiter, has a phalanx of moons was right. We have discovered sixty-two moons orbiting the planet to date, and they hold the key to understanding why the rings have such an intricate structure. Saturn is like a mini solar system, with the moons orbiting like planets around the Sun. From Earth we can only see the larger moons, but as Pioneer, Voyager and Cassini explored the system in more detail, it has become clear that these sixty-two satellites are a weird and wonderful bunch.

Dione is typical of Saturn's icy moons. Discovered by Cassini in 1684, it looks similar to our own moon but its composition is very different. It's about two-thirds water but the surface temperature is -190 degrees Celsius, and at those temperatures, the surface behaves like solid rock. It is a world covered in craters and ice cliffs and, just like our moon, it's tidally locked to Saturn, which means the same side of Dione always faces towards its master.

Iapetus is the third largest moon of Saturn and also one of the most mysterious. It has a strange ridge running around its equator and is so far away from Saturn that if you were to stand on its surface, you would be able to see Saturn's entire ring system dominating the night sky. When Cassini discovered this moon in 1671, he couldn't understand why he could only observe it when it was on the western side of Saturn. When it should have been on the east, it simply disappeared from view. Cassini concluded, correctly, that the reason for this mysterious observation was the great contrast between the two sides of the moon: one half is clean ice and the other is coated in black dusty deposits. The strange split personality of Iapetus' surface has led to it being nicknamed the Yin and Yang moon. The Cassini spacecraft has explored Iapetus in intricate detail, and it is now believed that half of the surface is covered in a layer of carbon. These dark deposits are thought to have come from an initial meteorite impact, but since that impact another characteristic of Iapetus has made the dark and light contrast of the two sides even more pronounced. Iapetus has a very slow rotation rate; one day on this moon lasts seventy-nine Earth days. This means that each side of the moon spends a long time being warmed by the dim light of the distant Sun, and this in turn leads to the highest daytime temperatures and the lowest nighttime temperatures of any of Saturn's satellites. Like all dark surfaces, the carbon-covered ice on the dark side absorbs more heat than the bright, reflective ice on the light side. This extra heat causes the ice to evaporate, leaving behind a carbon residue that makes the dark side even darker (and as the evaporated water moves to the cooler side, it makes the light side lighter, too). The darker the surface gets, the more heat it absorbs, increasing the temperature further and so creating a runway effect that over millions of years has divided the surface of Iapetus neatly in two.

Saturn's largest moon is Titan. This giant moon is bigger than the planet Mercury, almost as big as Mars and is the second largest moon in the Solar System. The unique thing about Titan is its atmosphere: it is the only moon that we know of that has a fully developed atmosphere. And what an atmosphere it is! Four times as dense as our atmosphere here on Earth, it is rich in organic molecules may have a chemistry very similar to that of the primordial Earth, before life began. Titan is without doubt one of the most fascinating places in the Solar System, and we'll return to it in detail in Chapter 4.

Hyperion is a moon unlike any other. It is not round, and its battered surface has the texture of a sponge. One theory is that Hyperion is a comet that drifted in from the distant icy reaches of the Solar System and was captured by Saturn's gravity.

Saturn's moons are a truly breathtaking collection of diverse and fascinating objects, but they aren't just a celestial freak show – they are the driving force behind the beauty and structure of the rings ◉

DIONE

DISCOVERED: 1684
APPEARANCE: Similar to the Moon – it is covered in craters and made of one-third rock; two-thirds ice
DIAMETER: 1,123 kilometres (696 miles)

TITAN

DISCOVERED: 1655
APPEARANCE: Formed by Earth-like processes, it is craterless with probable areas of liquid methane
DIAMETER: 5,150 kilometres (3,193 miles)

IAPETUS

DISCOVERED: 1671
APPEARANCE: One-half is clean ice and the other coated in black dusty deposits
DIAMETER: Saturn's third largest moon; 1.471 kilometres (0.912 miles)

HYPERION

DISCOVERED: 1848
UNIQUE FEATURE: Potato-shaped
DIMENSIONS: 410 km x 260 km x 220 km (254 ft x 161 ft x 136 ft)

Perhaps the most remarkable of Saturn's moons is buried in the heart of the E ring. This moon, Enceladus, is rapidly becoming one of the most intriguing places in the Solar System.

RIGHT: Enceladus, Saturn's sixth largest moon, is an astronomical mystery. It has defied geological theory; somehow this tiny icy moon has survived when it should have long since died.

ENCELADUS: THE BRIGHTEST MOON

The icy moon of Enceladus was first discovered by Sir Frederick William Herschel on 28 August 1789. Herschel was trained as a musician and composed twenty-four symphonies in his lifetime, but it was his inventiveness as an astronomer that ultimately assured his place in history. Herschel's crowning achievement was the discovery of Uranus in 1781, which he originally named Georgium Sidus in honour of King George III, who was passionately interested in astronomy. Despite believing that every planet was inhabited, including the Sun, Herschel was a brilliant telescope builder, innovator and observer of the night sky. He founded something of an astronomical dynasty. His sister Caroline was also a brilliant observer, discovering several comets and nebulae, and his son, John, became a famous astronomer, too (see page 30). William Herschel lived to the ripe old age of eighty-four, a number that links him to Uranus in the most fitting of ways, as Uranus takes eighty-four years to complete its orbit of the Sun.

In 1789, Herschel built the largest and most famous of all his telescopes in the garden of his home in Slough. This twelve-metre (forty-foot) telescope was then the largest telescope in the world, and on the very first night Herschel used it he became the first person to see Saturn's sixth largest moon.

Enceladus is tiny in comparison to our own moon, and in diameter it is smaller than the length of United Kingdom. For 200 years after its discovery we learnt little more about Enceladus beyond Herschel's initial observations. Besides knowing that it was made of water ice, we knew only its orbit and had estimations for its mass and volume. There was one other intriguing property of Enceladus that marked out the tiny moon as an astronomical curiosity: Enceladus is the most reflective object in the Solar System; its icy surface reflects almost all the sunlight that strikes it. Over a billion kilometres

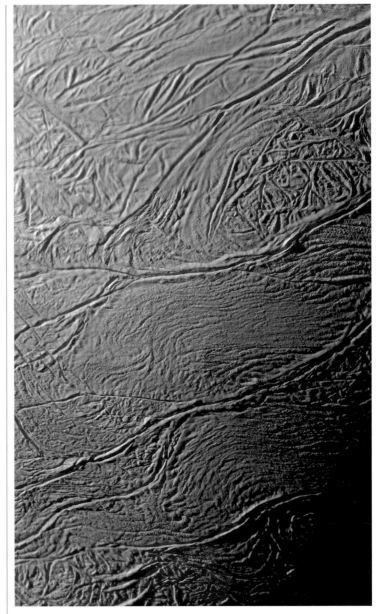

BELOW LEFT: The dramatic landscape of Iceland owes much to the separation of two major tectonic plates. As the continents slowly drift apart they tear apart the surface of the Earth.

RIGHT: The southern side of Enceladus is in direct contrast to the smooth surface elsewhere; here, four parallel trenches – the 'Tiger Stripes' – are carved into the ice.

away from Earth and only 500 kilometres (310 miles) across, however, the extraordinary secrets of Enceladus remained a mystery for over two centuries since there was no telescope large enough to uncover them. To do that, we had to fly there.

Our first proper glimpse of Enceladus came in 1981 when the Voyager spacecraft passed within 87,000 kilometres (54,000 miles) of the moon. Buried deep within the E ring of Saturn's intricate system, the images it took of Enceladus revealed something that no one was expecting. This ancient moon was not covered with impact craters as had been assumed, instead vast swathes of its surface were smooth. Virtually every moon in the Solar System is riddled with impacts from asteroids, so if the evidence of these destructive collisions has disappeared, then this demands an explanation. It is impossible for any place in the Solar System to have escaped the heavy bombardment of debris from space over billions of years, so the surface of Enceladus must be young. Its terrain must be constantly regenerating, erasing the record of countless thousands of collisions, so this frozen moon must be geologically active.

Only now that we have the images from Cassini have we begun to understand the strange and wonderful truth behind the smooth surface of Enceladus. Its heavily cratered northern hemisphere looks like any other icy moon, but the southern hemisphere tells a very different story. Its smooth, crater-free surface is scarred by canyons and riven by cracks and it looks remarkably similar to the geology of Earth, but carved in ice rather than rock. Right over the South Pole are the most extraordinary features we have found on Enceladus. An image photographed in incredibly high resolution by Cassini in July 2005 (opposite) shows four parallel trenches over 130 kilometres (80 miles) long, 40 kilometres (25 miles) apart and possibly hundreds of metres deep. Nicknamed the 'Tiger Stripes', these features look like tectonic fault lines found on our home planet, but Earth's geology is powered by the powerful heat source of its molten core. This heat is partly left over from Earth's formation 4.5 billion years ago, and partly due to the slow decay of heavy radioactive elements in the core. But a tiny world like Enceladus should have lost its meagre supplies of heat long ago to the cold of space, and should surely be geologically dead.

Iceland is home to some of the most dramatic landscapes on Earth. Sat on the dividing line between two continents, this great divide is one of the best places on the planet to explore the geological origin of Enceledus' mysterious Tiger Stripes. In this dramatic landscape you can see the inner workings of our planet exposed. The North American plate, the land mass that forms the western half of the North Atlantic and the United States, Canada and parts of Siberia,

is moving slowly west, whilst the Eurasian plate, comprising Europe and Northern Asia, is drifting east. Standing at the boundary, you can see the result of the inexorable drift of the continents, ripping apart the surface of the Earth and creating a plain of new crust between two towering cliff tops, formed from molten lava pushing up from deep beneath the surface. Carolyn Porco, head of the Cassini imaging team, believes something similar may be happening on Enceladus. Standing on the edge of the continental cliffs, she explained to me how such a rift valley could be created on tiny Enceladus, sculpted not from molten rock but from ice.

On 14 July 2005, Cassini flew directly over Enceladus' South Pole at a distance of just 175 kilometres (109 miles) from the surface. Using the infrared spectrometer built into the spacecraft, Porco and her team discovered the first direct evidence of geological activity beneath the surface of this moon. The thermal readings taken showed hot spots under the Tiger Stripes; the average surface temperature of Enceladus is around 75 Kelvin, but around the stripes the temperatures reached at least 130 Kelvin. This was a startling

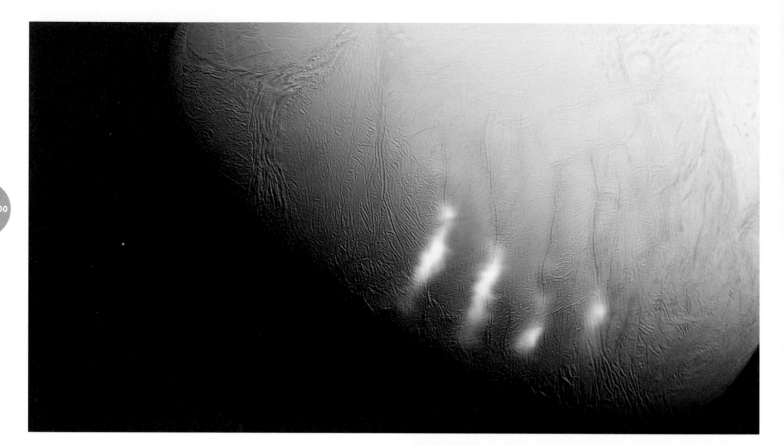

discovery. There is more heat coming out of the southern polar cap of Enceladus than is coming out of the equatorial regions. As Porco commented, it would be like finding that there's more heat coming out of Antarctica than the Equator here on Earth.

The greatest revelation, however, was yet to come. During November 2005, Cassini photographed Enceladus just as the sun was setting behind it (see right). What it showed has become one of the most remarkable discoveries ever made in the outer solar system. The backlit images revealed giant fountains erupting from the Tiger Stripes, volcanoes blasting out ice instead of rock. For Carolyn Porco these images were the culmination of a journey that began with her first work on the Voyager probe a quarter of a century before. 'Those images blew everybody away. I mean, that was like game over, you know? Here you have these dozen or more narrow jets and they just look ghostly and fantastic.'

Until a few years ago, Enceladus was thought to be an unremarkable world, a small, frozen, barren lump of rock and ice. But those fountains of ice erupting thousands of kilometres out into space reveal that there's something incredibly interesting going on beneath its surface. To understand exactly what's going on requires a journey to another one of Iceland's geological wonders: the Great Geysir in the Haukadalur Valley, Iceland. First documented in the Middle Ages, this regular towering fountain of boiling water and steam gave its name to the phenomena wherever it is found on Earth. The Great Geysir is currently dormant – it was last revived by an earthquake in 2000 – but just a few metres away is the Strokkur Geyser, which erupts every few minutes. Geysers are the earthly phenomena most like the ice fountains of Enceladus.

TOP: Fine ice particles and vapour spew out of the 'Tiger Stripes', which are believed to be the hottest part of Enceladus. This geological activity ensures the moon's survival.

ABOVE: This dramatic image, captured by Cassini, shows fountains of ice shooting out of Enceladus from the location of the Tiger Stripes.

BELOW: Geysers, towering fountains of boiling water and steam that erupt from below the Earth's surface, are the closest phenomena we have to the ice fountains of Enceladus.

'Those images blew everybody away. I mean, that was like game over, you know! Here you have these dozen or more narrow jets and they just look ghostly and fantastic.'

Geysers on Earth require three things: a ready source of water, an intense source of heat just below the surface and just the right geological plumbing. If the geysers on Enceladus are based on a similar mechanism, this raises an intriguing possibility. There must be a source of liquid water beneath the surface of the moon; small lakes or perhaps even an ocean that feeds the explosive volcanoes of ice. Yet Enceladus is a billion kilometres away from the Sun, in the cold outer reaches of the Solar System, and it is far too small to have retained any meaningful source of heat in its core. So where does that heat come from? On Earth, the geysers are driven by the same primordial heat source that powers the drift of the continents, but Enceladus is so tiny that its core should be frozen solid.

Enceladus must therefore be getting its heat from somewhere else. The source in all probability comes from its peculiar orbit around Saturn. Enceladus moves around Saturn in an elliptical orbit – in other words, its orbit is not a circle. This means that during each orbit Enceladus moves closer and then further away from Saturn. This eccentric orbit has a profound effect on Enceladus, changing the gravitational force exerted on the moon during every turn. As the difference in forces between the near and far sides of the moon changes, it literally flexes the moon as it travels around Saturn, distorting its shape and creating vast amounts of friction deep within. Friction causes heat,

and it is thought that the interior of Enceladus is heated just enough to melt a small underground ocean of water. As this water meets the vacuum of space, it immediately vaporises and explodes out of the surface, creating a true wonder of the Solar System.

Geysers on Earth are incredibly impressive natural phenomena, but they pale into insignificance when compared to the ice fountains of Enceladus. While on planets geysers erupt every few minutes at most, blasting boiling water twenty metres (sixty-five feet) into the air, on Enceladus the plumes are thought to be erupting constantly, and for them, the sky is the limit. Bursting through the surface at thirteen hundred kilometres (eight hundred miles) an hour, they rise thousands of kilometres into space. They must be one of the most impressive sights in the Solar System.

The water vapour in the plumes then freezes into tiny ice crystals. Some of it falls back onto Enceladus' surface, giving the moon its reflective icy sheen, but the rest keeps going all the way round Saturn. The ice fountains are creating one of Saturn's rings as we watch; the whole E ring is made from pieces of Enceladus.

Enceladus is not the only moon that shapes the rings, though; Saturn's other moons also play a crucial role in creating these beautiful patterns but they do so indirectly ◉

Bursting through the surface at thirteen hundred kilometres an hour, they rise thousands of kilometres into space. They must be one of the most impressive sights in the Solar System.

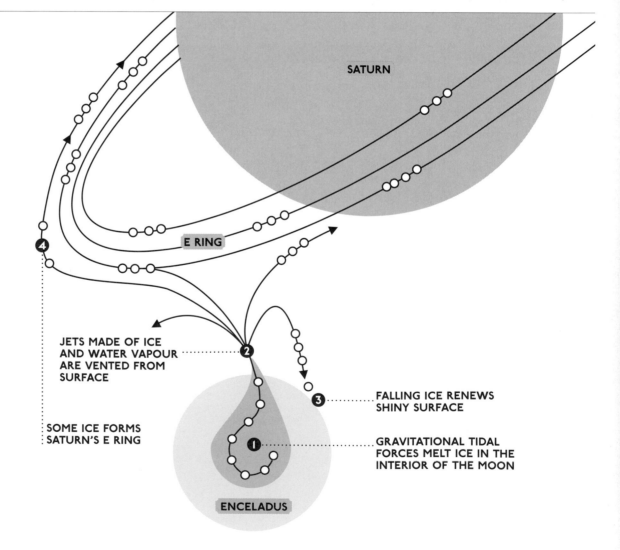

SATURN'S ICE RINGS
The constantly erupting ice fountains on Enceladus send out plumes of ice, which in turn replenish Saturn's E ring.

SATURN

E RING

JETS MADE OF ICE AND WATER VAPOUR ARE VENTED FROM SURFACE — ❷

❸ ········· **FALLING ICE RENEWS SHINY SURFACE**

❹

SOME ICE FORMS SATURN'S E RING

❶

········· **GRAVITATIONAL TIDAL FORCES MELT ICE IN THE INTERIOR OF THE MOON**

ENCELADUS

GRAVITY: THE GREATEST SCULPTURE

BELOW: The behaviour of the shifting sands in the Sahara desert provide an excellent model for explaining how moons form in Saturn's rings.

The Sahara desert may seem an unlikely place in which to explain Saturn's rings, but the behaviour of the sand in the desert can help us to understand how the moons form the patterns in their planet's rings.

At first sight the Sahara desert seems an immensely chaotic place; billions of grains of sand being blown randomly around by the desert winds. But look a little bit closer and you start to see order amidst the chaos. There are sand dunes as far as the eye can see, and remarkably the angles of the front of all the sand dunes are exactly the same – never exceeding thirty-four degrees. In the Sahara the emergence of that order is driven by the desert winds that always blow in the same direction, day after day, year after year, moving the sand around. The angle of the dunes is related to the physics of tumbling grains, and is a property not only of sand but of all small grains –

salt would behave in a similar way. Try pouring some salt into piles on the table at home – you'll see that the little pyramidal heaps all have the same angled sides. In nature, simple physical processes can create ordered structures that look for all the world as if they have been sculpted by a great artist. In the Saturnian system, the order, beauty and intricacy of the rings is created not by the desert winds roaring across the sand, but by the force of gravity.

ORBITAL RESONANCE

Gravity is a simple force to describe. For the purposes of understanding Saturn's rings, we need only use Isaac Newton's description of gravity first published in 1687. Albert Einstein completely rewrote our understanding of gravity in his theory of General Relativity, published in 1915, but his superior and more accurate theory is

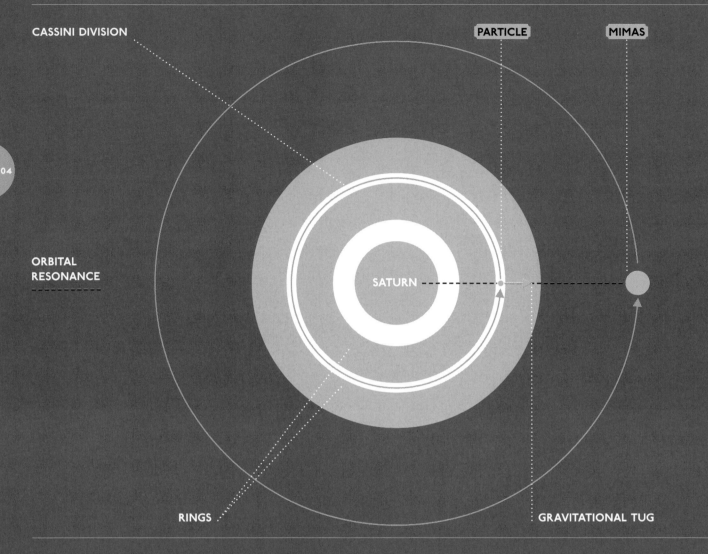

not necessary for understanding the clockwork of the Solar System, with the exception of the orbit of the planet Mercury, which is so close to the massive Sun that relativistic effects are important.

Gravity acts between all objects that have mass, providing an attractive force that is proportional to their masses and falls away with the square of the distance between them. (If we were using Einstein's theory, we should say energy as well, but we don't need such subtleties here!) So Saturn's moons gravitationally attract the tiny particles of ice in the rings towards them, and if you double the distance between the ice particle and the moon, the force of gravity drops by a factor of four. Imagine the complexity of the gravitational tugs on the ice particles in the rings as they orbit the giant planet, constantly being pulled by Saturn's vast phalanx of orbiting moons in an intricate complex dance. It is this complex gravitational field created by Saturn and its moons that gives rise to the structure in the rings.

This gravitational mishmash is present around any planet that has moons, but Saturn is special because there is a sheet of orbiting dust and ice sprinkled throughout the system, allowing us to see gravity in action. You may

have sprinkled iron fillings on a sheet of paper over a bar magnet and watched them line up into a beautiful pattern that reveals the hidden magnetic field. This is exactly what is happening around Saturn, except it is the gravitational field that is being revealed before our eyes by billions of tiny sprinkles of ice.

The easiest structure in the rings to explain is the biggest: that is the vast gap in the rings between the A and B rings, known as the Cassini division. This huge swathe of the rings is devoid of ice particles because of the influence of the moon Mimas – the Death Star moon – which orbits well outside the rings. It is a general feature of orbits that the further away from the planet the orbit is, the longer it takes to complete a single orbit. All the rings are inside the orbit of Mimas, and so all the particles orbit Saturn faster and constantly overtake Mimas. When they get to their closest approach to Mimas, the gravitational force exerted by the Moon on the particles will be at its greatest, disturbing their orbit a little. For most of the ring particles, this extra little kick happens at quite random points during the orbit, and the overall effect cancels out.

But there are particles in the rings whose orbits have an interesting relationship with Mimas. They go round

BELOW: One of Saturn's outer rings, the F ring, is twisted into a spiral as two moons, Pandora and Prometheus, pass close by and distort it by their gravitational pull.

Saturn twice for every single orbit of Mimas. This means that on a regular basis they will meet Mimas – they get close to the moon on one out of every two of their orbits, because by the time they have gone around Saturn twice, Mimas will have returned to the same position. This special relationship has a name – we say that the particle and Mimas are in an orbital resonance. In other words, the particles in resonance with Mimas are the ones at just the right distance from Saturn to keep meeting it periodically as they orbit around it. This means that they get periodic gravitational kicks on a regular basis, and their orbits are disturbed.

Perhaps you can guess what the effect will be: if a particle has an orbit around Saturn that is resonant with the orbit of Mimas, then its orbit will be changed by the regular gravitational kick and it will move out of that orbit, leaving a gap. This is the case for any particle that wanders into the Cassini division. The gap is the place in a space where ring particle orbits would be resonant with Mimas.

It is thought that much of the structure in Saturn's rings is down to resonances, some more complex than others, between the ring particles and one or more of the planet's moons.

There are other, more subtle but beautiful effects. As the moons orbit Saturn, we can see their gravitational effects sweeping through the rings. A series of images taken by Cassini reveal the moons as they work and show gravity in action. As the moons pass close to the rings, their gravitational pull tugs the ring particles towards them, distorting their shape. The F ring, one of the outer rings, is twisted into a spiral shape by two moons, Prometheus and Pandora. In the image above you can see how Prometheus drags plumes of material away as it passes close to Saturn's rings.

This exquisite structure, so delicately sculpted by the action of gravity, is an immense part of the wonder of Saturn's rings, because it is such a vivid illustration of how a simple force of Nature can carve order out of chaos.

But perhaps more than that, understanding how Saturn's moons shape the rings can shed light on the events that shaped the early Solar System; events that helped to create the world we live in; events that remained unknown to us until the greatest series of human expeditions in history opened our eyes to the chaotic origins of our solar system ◉

VIOLENT BEGINNINGS

It seems that between 4.1 and 3.8 billion years ago the Moon came under an extraordinary attack, bombarded in a meteorite storm that transformed and shaped the surface we see today. This showering of debris should also have affected the Earth and other inner planets. Many scientists now believe that this is evidence of an incredibly violent period in our solar system's history, known as the Late Heavy Bombardment. But what could have caused this colossal bombardment and turned the Solar System into a shooting gallery?

RIGHT: The Barringer Crater in Arizona, USA, was named after David Barringer who was the first to suggest that it was produced by meteorite impact.

THE LATE HEAVY BOMBARDMENT

On 26 July 1971, the Apollo 15 mission blasted off from the Kennedy Space Center in Florida with astronauts Scott, Worden and Irwin on board. This was the fourth Apollo team to land on the Moon and the first 'J-class mission' designed to put scientific investigation at the heart of the endeavour. On board the lunar module was an audacious piece of kit that transformed the capability of the astronauts to gather data. The Lunar Rover Vehicle, or 'moon buggy' as it became known, was a battery-powered car designed to allow astronauts to roam across the lunar surface and gather large samples of the rocks, allowing the geology of the Moon to be studied in much greater detail than ever before.

Landing in an area known as Mare Imbrium (Sea of Rains), the crew used the buggy to collect over 77 kilogrammes (170 pounds) of lunar rock samples. Mare Imbrium is a vast basin on the Moon's northwesterly surface that was created by a colossal impact early in its history. Its smooth surface was created when the then volcanically active moon flooded the impact crater with lava, creating the flat surface we see today. The moon buggy never returned to Earth and sits in Mare Imbrium to this day, but the samples it helped to collect did return and analysis of them has given us a unique and valuable glimpse into the violent early history of our solar system.

Many of the rock samples collected on the Apollo 15, 16 and 17 missions were impact melt rocks, created by the extreme conditions of direct meteorite impacts. Samples

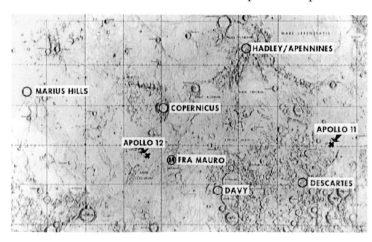

ABOVE: It is thanks to technology such as the moon buggy that astronauts and scientists can now navigate around the Moon and select landing areas in previously unexplored areas.

like this collected from all over the Moon have been dated using radioactive dating techniques and the results throw up a surprising pattern. A significant number of the rocks seem to have been created by impacts that took place in a relatively short timeframe during the Late Heavy Bombardment.

The Late Heavy Bombardment didn't only affect the lunar surface. If the Moon was showered in cosmic debris, the Earth and other inner planets should have suffered the same fate. Current estimates suggest that during this period Earth would have been peppered with thousands of impacts, with many creating impact craters over 1,000 kilometres (620 miles) across and some up to 5,000 kilometres (3,100 miles) in diameter. The cause of this bombardment may lie in exactly the same phenomenon that shaped the structure and complexity of Saturn's rings – orbital resonance.

Resonance can be much more than a delicate sculptor because it is not only small moons and particles of ice that can enter orbital resonance with each other. It is now thought that billions of years ago the two giants of the Solar System, Jupiter and Saturn, entered a resonance. In the case of Saturn's rings, resonances between ice particles in the Cassini division and the moon Mimas change the orbit of the particles, throwing them out of the gap. Likewise, if planets enter a resonance, the orbits of the planets are changed; and when planets start to fly around, the Solar System becomes an incredibly turbulent and violent place.

Using detailed simulations of the early Solar System, it is now thought that Saturn, Uranus and Jupiter formed much closer to the Sun than they are today. Their orbits drifted slowly for hundreds of millions of years until Jupiter and Saturn entered a resonance. Once every cycle, the two planets aligned in exactly the same spot, creating a gravitational surge that played havoc with the orbits of all the other planets. Saturn, Uranus and Neptune all migrated outwards and plunged the Solar System into a violently unstable era, triggering the Late Heavy Bombardment. In particular, Neptune was catapulted outwards and smashed into the ring of icy material in the outer solar system, randomly scattering them into orbits that crisscrossed it.

For a hundred million years, the Solar System turned into a shooting gallery as a rain of comets ploughed through it, peppering the planets and creating many of the craters we see on the planets and moons today. It's remarkable to think that this Late Heavy Bombardment, 3.6 billion years ago, was triggered by the same subtle phenomena, orbital resonance, which delicately sculpts Saturn's rings today ◉

BELOW: The Apollo Lunar Roving Vehicle (more affectionately and simply known as the moon buggy) revolutionised explorations of the Moon. This electric vehicle enabled astronauts to venture further across the Moon's surface to collect a wider range of samples.

A GIFT TO EARTH

Today, it's almost impossible to find any direct evidence on our planet of the Late Heavy Bombardment, the impact craters long ago became shrouded by the ever-changing surface of the Earth. However, one defining characteristic of our planet that we can observe may be a direct result of the thousands of comet impacts that battered the Earth around 3.6 billion years ago.

Comets have a different composition to asteroids. This becomes visible when they venture into the inner solar system, close enough to be warmed by the Sun. As they absorb the Sun's heat they display a tail and an atmosphere as the ingredient they have in abundance evaporates away into space: water.

It is thought that the intense onslaught of comets during this turbulent time changed the Earth's environment radically and dramatically, but those changes weren't necessarily catastrophic. As the Solar System descended into chaos it seems that that a significant amount of the water in the Earth's oceans today was delivered by the impacts of water-rich comets and other objects during the Late Heavy Bombardment, which means that impacts could have played a key role in the development of life on Earth.

Before the Late Heavy Bombardment the Earth may have been a relatively barren rock, starved of water; afterwards it supported the oceans that would become the crucible for life. Without this water delivered in the Late Heavy Bombardment, life on Earth may never have evolved. It's quite a thought that this water-rich world we see today may have been shaped by violent resonances generated by the orbiting gas giants Jupiter and Saturn.

It is one of the most wonderful gifts to the astronomer that the finest example of the remarkable journey from chaotic collapsing dust cloud to delicately sculpted beauty is also the most stunning laboratory for studying how the Solar System works: Saturn's rings.

It's often the case in science that answers to the most profound questions can come from the most unexpected of places. Saturn's rings were initially studied because of their beauty, but understanding their formation and evolution has led to a deep understanding of how form, beauty and order can emerge from violence and chaos.

It is also worth remembering that life on Earth is a part of the Solar System. We are ordered structures, formed from the chaos of the primordial dust cloud 4.5 billion years ago. We are just as much a product of gravitational collapse as the rocky inner planets, the majestic gas giants and the impossibly delicate artistry of Saturn's rings. We were formed by the same laws of Nature, and important steps in our formation are written in the sky for us to read. We are part of the heavens and intimately connected to them. And that is truly one of the wonders of the Solar System ◉

CHAPTER 4

THE THIN BLUE LINE

EXPLORING EARTH'S ATMOSPHERE

The Solar System is a violent, inhospitable place. On every planet and moon we've explored, from our nearest neighbours to the most distant corners of the Sun's realm, we have encountered extremes. We've discovered worlds defined by the fiercest heat and the bitterest cold; seen landscapes sculpted by overwhelming pressure, and witnessed storms the size of planets. Amongst all these hostile wonders sits our Earth, an oasis of calm amidst the violence of the Solar System. Yet all that separates us from what's out there, from the extremes that sit above our heads, is a thin, flimsy envelope of gas. Our atmosphere may be an invisible presence in our daily lives but it's thanks to this thin blue line that we have the air that we breathe, the water that we drink and the landscape that surrounds us.

JOURNEY TO
THE EDGE OF
THE EARTH

Hidden away on the outskirts of Cape Town International Airport is an aeronautical toy box like no other. From Blackburn Buccaneers to Hawker Hunters, this place is home to the finest and largest collection of classic military aircraft in the world – and they are all still flying. One aircraft in the collection, however, stands out as a truly wondrous piece of engineering. The English Electric Lightning is a supersonic jetfighter that was designed and built in the 1950s. Capable of flying at over twice the speed of sound (Mach 2.27, 2,400 kilometres or 1,500 miles per hour), this beautiful machine was used by the Royal Air Force for almost thirty years as an interceptor aircraft – designed to hunt down and destroy enemy bombers at great speed.

As well as its swiftness, the Lightning is also renowned for another characteristic: it can fly incredibly high. Although it was officially a military secret, it's now well documented that the Lightning can fly way beyond its designated operating height of over 18,000 metres (60,000 feet). In 1984, during a NATO exercise, an RAF pilot took the craft to 27,000 metres (88,000 feet) to test its ability to intercept the supposedly untouchable U2 spy plane. It not only succeeded in carrying out its mission but it also carried the pilot to a height that took him above 99 per cent of the Earth's precious atmosphere.

I grew up with Lightnings. In the 1970s the Lightning was the aircraft every plane-spotting kid loved – a piece of science fiction that would not have looked out of place in *Star Wars*. It was the fastest, sleekest and most powerful interceptor on the planet. Up close it isn't a delicate or balletic aircraft. Anything that flies at twice the speed of sound has to be solid; no rattles or creaks. The cockpit is small and surprisingly high off the ground. You feel perched out on a limb, bolted securely

In the 1970s the Lightning was the aircraft every plane-spotting kid loved – a piece of science fiction that would not have looked out of place in Star Wars.

but precariously onto two Rolls-Royce Avon engines and tanks of combustible gases and fluids. White instrumented dials in grey boxes and toggle switches labelled with a cold-war-era font are randomly slotted around the ejector seats. Between your legs (which are attached by seat-belt fabric to the automatic leg retraction system to preserve your knees, should you decide to kick out), is the control stick, barnacle-encrusted with gun-triggers and missile launch controls. It is, in short, a place that any kid with a bit of bottle would want to be.

Engine start-up in a jet fighter is always a careful affair. The pilot watches the dials, looking for any abnormality in temperature. This is analogue, and the information about the health of the Avons is in the nuances of the needles. It is also surprisingly quiet and vibration-free from the inside – more airliner than war machine. When the pilot is happy, ex-RAF Lightning XS 451 gently taxis to the end of the runway at Cape Town International Airport, behind a South African Airways Airbus A340.

I was waiting for a brutal start to the take-off roll, but the pilot accelerates the Lightning quite gently along the runway. It feels no faster than a passenger jet – until we get airborne. I am then treated quite unexpectedly to the Lightning's party trick; a rotation take-off. The afterburners are kicked in as soon as the undercarriage is retracted, and the aircraft enters a near-vertical climb, rolling to one side as it goes to reduce the stress on the airframe. It is a rocket launch.

Only fifteen minutes after take-off I am reminded why we came here to film with the silver machine. We reach 17,700 metres (58,000 feet), inverted to reduce airframe stress in the climb, then flip over into level flight 6,000 metres (20,000 feet) higher than a passenger jet. In an instant, an expansive and powerfully moving vista materialises. I see the Earth, but not as the geometrically flat expanse of land I'm used to gazing across from an airliner window. It is curved. Very curved. Overwhelmingly small, because there is enough curvature to allow the mind to recreate the rest and construct a tiny planet. It is at once majestic and diminutive.

I'm seeing the land, the air and the beginnings of the vacuum of infinite space in a single field of view, and the word to describe it is 'fragile'.

Stretching upwards from the horizon is a graduated wash of fading colour; bright sky blue where the land meets the air, but quickly darkening towards a deeper, twilight azure. I'm seeing the land, the air and the beginnings of the vacuum of infinite space in a single field of view, and the word to describe it is 'fragile'. Most excellent canopy, indeed, but a surprisingly delicate majestical roof. This is the thin blue line that shields us from infinity, and strangely you have to climb above ninety per cent of it to be able to see and understand it.

We touch down forty minutes after take-off, having journeyed to the edge of planet Earth and back. For me the Lightning has changed its character. The *Star Wars* interceptor has become an enabler; a necessary tool to deliver a vital experience. It would not be possible to see our planet and atmosphere wandering delicate and precious through the emptiness without the brutal power of the cold-war interceptor. Engineering is the route to enlightenment because it transports us to the places where our terrestrial perspective is shaken to breaking point and forcibly replaced. We are a complacent and parochial bunch, scuttling around on our rock beneath our foul and pestilent congregation of vapours, and if it takes a pair of afterburning Rolls-Royce Avons to free our minds, then so be it.

THE THIN BLUE LINE

The Earth would not be the wonderfully diverse place that it is without the thin blue line. It acts as a soothing blanket that traps the warmth of the Sun, yet protects us from the harshness of its radiation. Its movements can be traced in the gentlest breeze and the most devastating hurricane. The oxygen, water and carbon dioxide the atmosphere holds plays a fundamental role in the ongoing survival of millions of different species living on the planet. In this chapter we'll explore how the laws of physics that created our unique atmosphere are the same laws that created many diverse and different atmospheres across the Solar System.

When perfectly balanced, a world as familiar and beautiful as the Earth can evolve beneath the clouds, but the slightest changes can lead to alien and violent worlds. There are planets in our solar system that have been transformed into hellish worlds by nothing more than the gases in their atmosphere. Just as atmospheres can choke a planet to death, they're also powerful enough to shape their surfaces. There are even worlds out there that are all atmosphere – giant balls of churning gas where storms three times the size of the Earth have raged for hundreds of years. All atmospheres in the Solar System are unique, but the ingredients and forces that shape them are universal. At the heart of each is a fundamental force of nature that holds the Solar System together: gravity ◉

LEFT: The Hubble space telescope was launched by NASA in 1990. As the Earth's atmosphere blocks some light from space, by placing this telescope above it, we can receive clearer images of space.

THE BINDING FORCE OF GRAVITY

Gravity is by far the weakest known force in the Universe. You can see that because it's really easy for me to pick a rock up off the ground, even though there's a whole planet – Earth – pulling the rock down. I can just lift it up; incredibly weak, but important because it's the only force there is to hold an atmosphere to the surface of a planet.

Gravity is one of the four fundamental forces of Nature. Isaac Newton first explained it in 1687, describing it as the force by which objects with mass are attracted to one another. We now know that this is an approximation: Albert Einstein provided us with a more sophisticated picture of gravity in his General Theory of Relativity in 1915. Gravity exists because space and time are curved by the presence of mass and energy, and we feel the results of this curvature as an attractive force between all objects. For our purposes, and indeed for most purposes, Newton's simpler view is sufficient.

Compared to the other three fundamental forces of Nature – the strong nuclear force, the weak nuclear force and the electromagnetism – gravity is the weakest, and yet when combined with these forces, it creates the conditions for stars to ignite, holds planets in moons in their orbits and binds the galaxies together.

According to Newton, the force of gravity between two objects can be described by the simplest of equations. (see diagram opposite). With the help of G – the universal gravitational constant – the force between two objects is calculated by multiplying the mass of each and then dividing that by the square of the distance that separates them. This beautifully simple equation allows us to explain and predict so much about our universe and solar system. There is possibly no greater example of the power of simple mathematics than the story behind the discovery of the most distant planet from the Sun in our solar system – a planet whose existence was predicted by Newton's law of gravitation long before it was directly observed by the human eye.

Over four billion kilometres (three billion miles) away from Earth, the beautiful blue planet of Neptune is one of the coldest places in the Solar System and the only planet entirely invisible to the naked eye. Impossible to detect before the invention of the telescope, Neptune was probably first observed by Galileo in 1612, but it was mistaken for a fixed blue star in the night sky. This is because what Galileo didn't know as he observed it was that Neptune only appeared to be stationary, as on that evening this giant ice planet was beginning a retrograde loop (see page 76) and so was just changing

NEWTON'S LAW OF UNIVERSAL GRAVITATION

The gravitational force exerted by one mass on another is proportional to the product of the masses and inversely proportional to the square of the distance, r, between them. The magnitudes of the forces exerted by each object on the other, FI and F2, will always be equal in magnitude and opposite in direction. G is the gravitational constant.

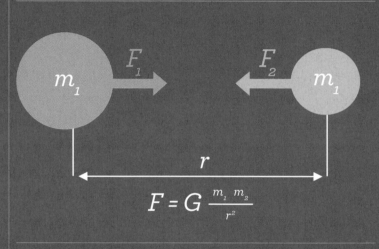

$$F = G \frac{m_1 \, m_2}{r^2}$$

direction in the night sky. The real clue to Neptune's existence, however, came not from the observation of this planet but from the observation of its nearest neighbour, Uranus.

Discovered in 1781 by William Herschel, Uranus was the first planet to be discovered since ancient times, so its passage through the night sky was plotted by hundreds of astronomers keen to follow this newest addition to the Solar System. In 1821, the first astronomical tables of Uranus' orbit were published by French astronomer Alexis Bouvarde. Using Newton's Law of Gravitation, the tables provided accurate predictions for the future position of the planet as it travelled around the Sun. It soon became clear, however, that Uranus was not behaving exactly as predicted. The path it took on its orbit did not agree with the path that Newton's Law predicted. At times on its orbit Neptune was either forward or behind its predicted position. Something seemed to be wrong. Astronomers of the time were baffled by the discrepancy and struggled for an explanation. Could Newton's Law be wrong? Or was the quality of the observed data at fault? The only other option seemed to be that something was disturbing the journey of this giant planet around the Sun. Many of these scientists thought the last of these reasons was the most likely, and that the disturbance to the orbit was due to a gravitational perturbation caused by an as yet undiscovered planet. If

BOTTOM: The farthest planet in our solar system, Neptune is also one of the coldest. It was probably discovered in 1612, but it was in 1989 that Voyager 2 became the first spacecraft to observe the planet. Neptune owes its distinctive blue colour to the methane in its upper atmosphere.

GRAVITATIONAL PERTURBATION

Here two planets are orbiting a common star. When the planets are at A, the gravitational force exerted by the outer planet on the inner planet causes the inner planet to accelerate, moving ahead of the position calculated by considering the sun's gravity alone. When the planets are at B, the reverse is true and the inner planet is decelerated. This slight deviation in the path taken by the inner planet is said to be due to a gravitational perturbation. This led to the prediction and discovery of the planet Neptune.

B

A

SUN

there was an unknown planet orbiting outside Uranus, then according to Newton there would be an additional gravitational force between this mystery planet and Uranus. This would alter Uranus' orbit, and explain the discrepancy between the theoretical predictions and the experimental observations.

In 1845–46, the French astronomer Urbain le Verrier and English astronomer John Adams independently calculated the mass and position of such a new planet. By using the observed data from Uranus, they could employ Newton's equation to calculate the mass and distance of this planet from the Sun and Uranus, according to the gravitational force it appeared to exert. Although it is uncertain who actually reached the end of the calculations first, their collective work led to the precise prediction of the existence of a giant planet orbiting outside Uranus. This prediction was rapidly confirmed when German astronomer Johann Galle made the first telescopic observation of Neptune in September 1846.

The story of the discovery of Neptune is a beautiful example of the predictive power of the Laws of Physics and the universal influence of gravity. In this case, it described the massive force between two giant planets orbiting the Sun, but it can also predict far more subtle interactions and explain the tenuous connection that joins a planet with its most nebulous characteristic – the atmosphere ◉

LOBSTERS ON THE OCEAN FLOOR

Our atmosphere – the thin, fragile layer of gas that surrounds Earth – is held to our planet by nothing more than gravity. Just as two planets exert a force on one another, so there is a force between each atom in our atmosphere and the Earth. This incredibly weak force binds these life-giving atoms to our planet. Whether it's oxygen, nitrogen, argon, carbon dioxide or any of the other gases that fill our sky, each atom is tenuously bound to Earth by the gravitational force between its tiny mass and the enormous bulk of our planet. This is all there is to prevent Earth's atmosphere disappearing into space. The more massive the planet, the greater the gravitational force binding the atoms in the atmosphere to the surface. Fortunately for us, the Earth has enough mass to keep a tight grip on the heavier gas molecules that make up our atmosphere. It holds them against the surface and allows almost every living thing to survive.

In the case of our planet, gravity has glued a cocktail of gases to the Earth; the main constituent being nitrogen. Almost seventy-eight per cent of the air around us contains this invisible, odourless and tasteless gas. One of the main reasons for this is that nitrogen gas is extremely stable and so reacts with very little, making it an incredibly long-lived gas in the atmosphere. Almost all of the rest of our atmosphere is made up of oxygen. At just twenty-one per cent of the total volume, it is by far the most abundant component after nitrogen. The inert gas argon is the third most abundant gas, at just under 1 per cent. The rest of the atmosphere consists of trace gases, such as carbon dioxide, neon, nitrous oxide and methane. These trace gases are so sparse that when added together they comprise just 0.039 per cent of the Earth's atmosphere.

We don't normally notice the presence of this vast mass of gas that surrounds us, but in fact we have evolved to live under a massive weight of air. There are five million billion tonnes of air surrounding the Earth, and at any one moment that huge mass of air is pressing on each and every one of us. Without even realising it, we all live our lives under pressure. On every square centimetre of our bodies there is a force that is equivalent to the weight of a one-kilogramme object pressing down. To put it another way, if the average person is about a metre square in area, the atmospheric pressure is the equivalent of a ten-tonne object pushing us.

Life on the surface of this planet survives surrounded by this enormous mass of gas, and just like lobsters scuttling around on the ocean floor, we have evolved to deal with the

This thin blue line makes the Earth the wonderfully diverse place that it is. It acts as a soothing blanket that traps the warmth of the Sun, yet protects us from the harshness of its radiation.

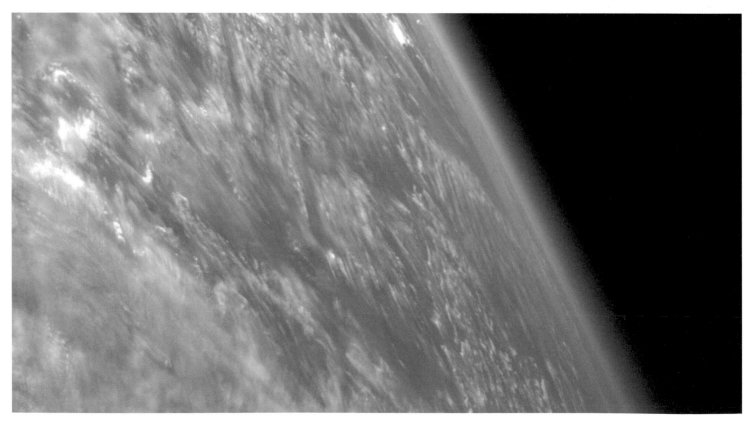

OPPOSITE: Here, over north-western Africa, we can clearly see the thin blue line that represents the Earth's atmosphere. This fine, fragile layer of gas follows the curvature of the Earth's surface, ensuring our survival on this planet.

Composed of 78 per cent nitrogen, 21 per cent oxygen and 1 per cent other constituents, the Earth's atmosphere acts as a shield against nearly all harmful radiation coming from the Sun and other stars, while trapping their warmth at a beneficial level.

BELOW: This spectacular picture was taken by the STS-125 crew of Atlantis as they returned home from Hubble in May 2009. Beyond the payload bay of the shuttle, the thin blue line of the Earth's atmosphere cuts through the blackness of space.

123

pressure so effectively, we don't even notice it's there. We just breathe it in and use the oxygen to allow our bodies to function. But that's not the end of our intimate relationship with the air around us; our atmosphere does more than just allow us to breathe, it protects us from the most powerful force in the Solar System: our sun.

UNDERSTANDING AIR PRESSURE

Air pressure is a slightly counter-intuitive concept. It's an easy linguistic mistake to make to say that all that vast mass of air above our heads is pressing down on us, crushing us onto the surface of the Earth. But that's not how air pressure works. It presses on us in every direction at once – otherwise how could we possibly be strong enough to support the equivalent of a ten-tonne object pressing 'down' on our bodies?

The air pressure is due to the billions of molecules in the atmosphere bouncing off us as they jiggle around. Imagine someone smashing a tennis ball into your face. As the ball bounces off, it hurts because the change in direction of the ball requires a force to act. Your nose provides that force, and since to every action there is an equal and opposite reaction, you'll feel that force on your nose! The molecules in the atmosphere are exactly the same as little tennis balls, only much smaller, so as they continually bounce off your body they exert a force on you. Pressure is defined as the force per unit area – in other words, it is the net effect of all the molecules of air bouncing off every square centimetre of your body. Thinking of it in these terms, it's easier to see that it doesn't matter whether the square centimetre of your body in question is pointing up, down or sideways – the number of air molecules that bounce off it will be the same and so the air pressure will act equally in all directions.

If you still don't believe this explanation, you can do this simple experiment. Half-fill a glass with water and carefully place a piece of paper over the top of the glass. Holding the paper in place, you can turn the glass over and then let go of the paper, and the pressure of the atmosphere pushing upwards on the paper will hold the water in the glass.

Our bodies are actually completely open to the air – there are no sealed air pockets inside us. This means that we can exist quite happily at much higher or lower pressures than atmospheric pressure. A scuba diver can happily descend to twenty, thirty or even forty metres below the surface of the ocean without any special equipment. At a depth of forty metres the pressure is five times the atmospheric pressure – that's equivalent to a fifty-tonne object pressing on every square metre of the diver's body! As long as the diver keeps breathing and popping her ears, the pressure inside and outside her body will remain in perfect balance and therefore she will feel no ill effects ◉

EARTH'S AMBIENT TEMPERATURES

The average temperature on Earth is a balmy thirteen degrees Celsius, but of course it varies enormously across the planet. The highest recorded temperature on our planet is 56.7 degrees Celsius in the Libyan deserts; the coldest is -89 degrees Celsius in the depths of Antarctica, but compared to other places in the Solar System our temperature swings are fairly gentle. The cause of this stability and the reason for our average temperature may appear to be straightforward. We are 150 million kilometres (93 million miles) away from the Sun and the distance from this heat source sets the amount of energy being received, which determines the temperature. Just as we expect to be warmer when we are closer to a fire, so it would be reasonable to expect that every planet gets warmer the closer it is to the Sun. But things aren't quite that simple.

RIGHT: Namib Desert, Namibia.

A TALE OF TWO ATMOSPHERES

As the Sun sinks below the horizon in the Namib desert in Namibia, south-western Africa, the temperature change from day to night can be as much as thirty degrees Celsius. That's an immense amount in just a few hours; one of the biggest day-to-night swings on the planet. The reason for this dramatic change is that the Namib desert is also one of the driest places on Earth.

The levels of water vapour in the Earth's atmosphere vary across the planet, but wherever there are small amounts of water vapour in the atmosphere there are large day-to-night temperature swings. This is because the ability of the atmosphere to trap heat is directly related to the insulating effect of the water vapour. In the dry environment of the desert the level of insulation is low, and so when the Sun disappears, the heat disappears quickly into space. There are many other gases in the atmosphere as well as water that act as insulators, turning our atmosphere into a warming blanket. These greenhouse gases, such as carbon dioxide, methane and nitrous oxide, trap the heat of the Sun, ironing out the difference between day and night so effectively that we think a change of thirty degrees is significant!

Difficult to see from Earth, Mercury suffers the biggest temperature swings of all the planets. This is because it has been stripped of the one thing that could protect it: its atmosphere.

BOTTOM LEFT: Here in the Namib desert, the change in temperature from day to night can be up to thirty degrees Celsius. This feels immense, but it is nothing compared to the extreme temperature change that occurs on planets such as Mercury.

BELOW: Mercury, the smallest planet in our solar system, has been stripped of its atmosphere since its inception. Left unprotected in this way and without insulation, Mercury experiences the largest temperature swings of all the planets.

HOW MERCURY LOST ITS ATMOSPHERE

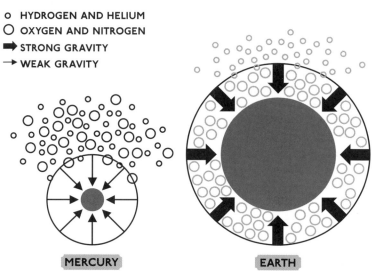

- ○ HYDROGEN AND HELIUM
- ◯ OXYGEN AND NITROGEN
- ➡ STRONG GRAVITY
- → WEAK GRAVITY

MERCURY   EARTH

THE GREENHOUSE EFFECT

Almost 100 million kilometres (60 million miles) closer to the Sun than the Earth is a planet where the temperature shift from day to night is immense. Mercury, the smallest of all the planets, is roughly fifty-eight million kilometres (thirty-six million miles) from the burning heart of our solar system. Difficult to see from Earth because of its proximity to the Sun, this tortured piece of rock suffers the biggest temperature swings of all the planets, from 427 degrees Celsius in the day to -173 degrees Celsius at night. This is all because Mercury has been stripped of the one thing that could protect it: its atmosphere.

Like all the rocky inner planets of the Solar System, Mercury had an atmosphere at formation. In fact it's thought all eight of the Sun's planets had similar atmospheres when they formed over four billion years ago, composed of lighter gases like hydrogen and helium, with smaller amounts of heavier gases like oxygen and nitrogen.

Planets hang on to their atmosphere by the force of gravity – it's the only way they can stop the fragile line of gas disappearing off into space – so the more massive the planet, the stronger the gravitational pull and the easier it is for the planet to keep hold of its atmosphere. The temperature of the atmosphere also affects this balance, because the hotter the atmosphere, the faster the molecules are whizzing around and the harder it is for the gravitational force to hang on to them.

The giant planets of the outer solar system, Jupiter Saturn, Uranus and Neptune, were large enough and cold enough to exert the massive gravitational force needed to hold on to the lighter gases such as hydrogen and helium, but on the inner, warmer and smaller rocky planets the story was very different. The lightest gases would have gradually escaped into space from Mercury, Venus, Earth and Mars, leaving behind atmospheres rich in heavier gases such as oxygen and nitrogen. Fortunately for us, the Earth is big enough and far enough away from the Sun to exert a force of gravity that can tightly hold on to these gases, and our atmosphere has been able to evolve with them in place over billions of years. On Mercury, the story is very different. Mercury is tiny compared to Earth; with a trip around its equator of 15,329 kilometres (9,504 miles), the surface area of Mercury is one-seventh of Earth's surface, and its mass is just five per cent of that of our planet. Coupled with its high surface temperature, that means the gravitational force is not strong enough to hold on to the heavier gases, so Mercury rapidly lost almost its entire atmosphere.

The impact of this on the two planets today is striking. Here on Earth, at sea level, in a volume about the size of a sugar cube there are twenty-five billion billion molecules of gas. On Mercury, in the same volume, there would be around a hundred thousand – over 100 million million times less. So Mercury was just too small and too hot to hang on to its atmosphere and the consequences for the planet were devastating. Atmospheres may be just a thin strip of molecules but they're a planet's first line of defence. Without them, a planet like Mercury is at the mercy of our violent solar system ◉

SURFACE TEMPERATURE

Although you would expect the surface temperature of a planet to decrease the further away it is from the Sun, the interaction between the Sun and the atmosphere means that some planets are warmer than they should be, such as Venus, whilst others are colder, such as the Earth.

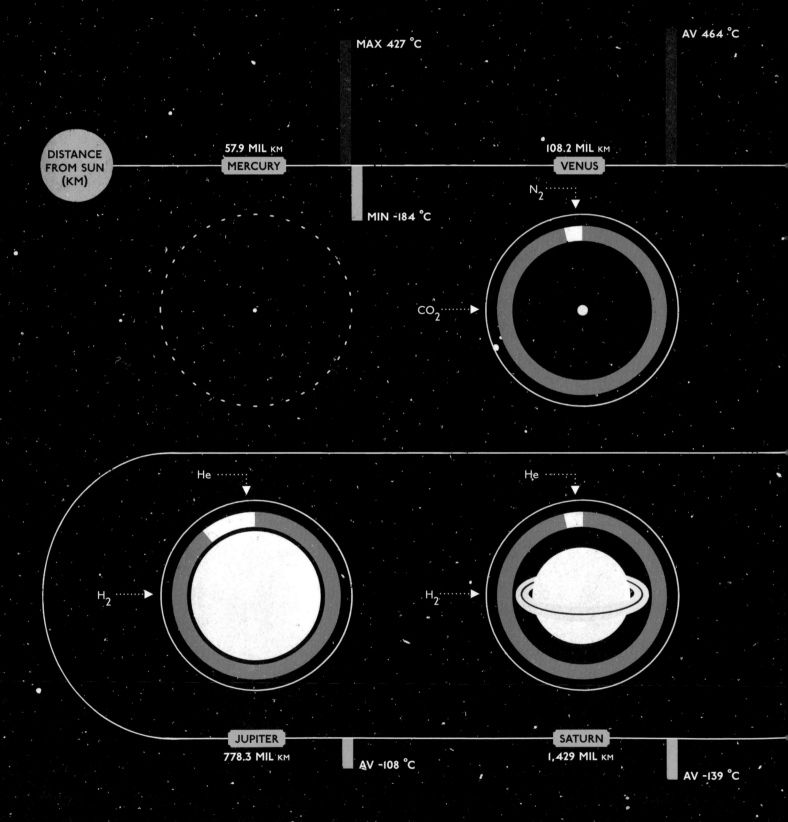

MAX 427 °C

AV 464 °C

DISTANCE FROM SUN (KM)

57.9 MIL KM
MERCURY

108.2 MIL KM
VENUS

MIN -184 °C

N_2

CO_2

He

He

H_2

H_2

JUPITER
778.3 MIL KM

AV -108 °C

SATURN
1,429 MIL KM

AV -139 °C

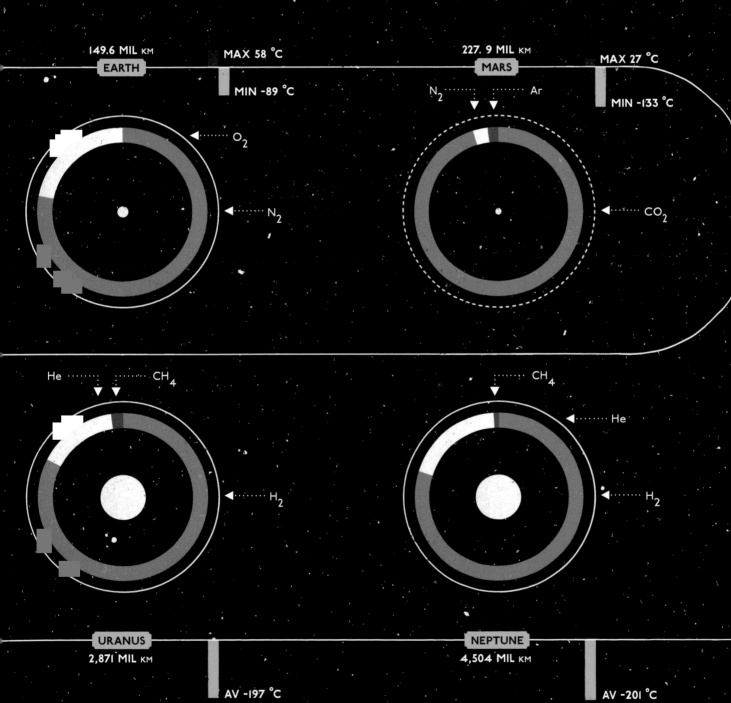

FULL ATMOSPHERE

PARTIAL ATMOSPHERE

WEAK ATMOSPHERE /
NOT WELL KNOWN

N_2 NITROGEN

Ar ARGON

He HELIUM

H_2 HYDROGEN

CO_2 CARBON DIOXIDE

O_2 OXYGEN

149.6 MIL KM
EARTH

MAX 58 °C

MIN -89 °C

O_2

N_2

227. 9 MIL KM
MARS

MAX 27 °C

MIN -133 °C

N_2 ········ Ar

CO_2

He ········ CH_4

H_2

CH_4

He

H_2

URANUS

2,871 MIL KM

AV -197 °C

NEPTUNE

4,504 MIL KM

AV -201 °C

FIRST LINE OF DEFENCE

BELOW AND BOTTOM: On 20 November 2008, the dark winter skies of western Canada were unexpectedly illuminated by a fireball five times as bright as the Moon. Saskatchewan was bathed in an eerie blue light as an asteroid entered the Earth's atmosphere.

The province of Saskatchewan in western Canada is a cold dark place to be in winter, but on 20 November 2008 the night sky was lit up by a fireball five times as bright as a full moon. The light show witnessed that night was the result of an asteroid – a space rock weighing about ten tonnes – entering Earth's atmosphere and landing in a place called Buzzard Coulee. It's certainly not unusual for rocks this size to hit the Earth (on average, it happens about once a month), but what was unusual about the Buzard Coulee meteorite was that its trajectory took it over quite densely populated areas so that tens of thousands, if not hundreds of thousands, of people saw and heard it. Most spectacularly, it was also captured by a lot of CCTV cameras charting its journey across hundreds of kilometres of sky. These images created a remarkable record of the meteorite as it streaked across the night sky at twenty kilometres (twelve miles) per second, turning it blue.

These remarkable images enabled a group of scientists to triangulate the impact site of the meteorite with far greater accuracy than is normally possible. Thus a team of meteorite hunters could search for the debris from the explosion in a precise location, a field just outside the city of Lloydminster.

Leading the team is University of Calgary professor Dr Alan Hilderbrand, one of the world's leading meteorite experts and a member of the team behind the discovery of the ancient Chicxulub Crater in the Yucatan peninsula of Mexico.

this in turn heats the meteorite until it is white-hot. For a brief time, this billion-watt bulb shone high in the sky; then, after just five seconds, its billion-year journey suddenly and dramatically comes to an end as it disintegrated in a series of explosions, peppering the fields below with lumps of rock the size of golf balls.

In just a few seconds, this survivor from the distant past became just another part of planet Earth. It must have been quite incredible to be standing in Saskatchewan on that night as the sky turned blue, watching these heavy rocks raining down.

Hunting for meteorite fragments requires a certain kind of talent. These rocks have streamed through the atmosphere with such intensity that their surface melts as it reaches temperatures of 6,000 degrees Celsius – the surface temperature of the Sun. This searing heat creates a tell-tale dark crust over the meteorite, so I was told we were looking for an oddly sculpted dark rock on the ground. (A rock that Hildebrand helpfully points out looks remarkably like the cow pats that litter this landscape.) However, the mind-numbing, slow search feels well worth it when I did find one of these astonishing rocks.

Each of these little rocks scattered across the frozen fields has had an amazing history. They would have approached the Earth as part of a meteorite known as a Chondrite. Chondrites are very old, having formed at the very beginning of the Solar System over 4.5 billion years ago. They have never been a part of a larger planet or moon; they are pure, wandering fossils of an earlier age.

If the meteorite had hit the ground intact, the explosion would have left a crater twenty metres (sixty-five feet) wide. Our planet was only spared this colossal impact because of the tenuous strip of gases that surrounds us, a strip of gas that we hardly notice. Nowhere is the crucial role of this protective blanket demonstrated more starkly then on the battered surface of our smallest neighbour – Mercury ◉

This vast crater, over 180 kilometres (112 miles) in diameter, is thought to date back to the end of the cretaceous period 65 million years ago, and it remains the prime candidate for the catastrophic event that wiped out the dinosaurs.

Fortunately for the people of Canada, Dr Hildebrand's latest search is on a far smaller scale; however, there is a still a huge amount of information to be gleaned from this lightweight impact. A ten-tonne rock travelling at fifty times the speed of sound has an extremely large amount of energy. 'It would be like stocking up 400 tonnes of TNT to explode,' explains Hildebrand. 'It's really quite dramatic.'

There is only one thing that can possibly protect the Earth's surface from a projectile with that amount of energy on a direct collision course with our planet; just one thing that can cause it to slow down and break up and so prevent all of the energy slamming into the Earth's surface in one place. Night after night our atmosphere slows down and breaks up meteorites just like this one. Entering the atmosphere at twenty kilometres a second and heading directly towards Buzard Coulee from the depths of space, this lump of rock and iron that was originally about the size of a desk, would have immediately begun to compress the thickening atmospheric gases in front of it. When air is compressed, it heats up, and

MERCURY'S CRATERS

On 30 January 2008, the NASA Messenger space probe took images of Mercury's surface. This was the first time humans had seen these parts of Mercury's scorched surface and it confirmed what had long been known about this tortured planet: it is a battered, barren piece of rock.

Until Messenger swept past Mercury, we knew little about the precise surface details of the planet. The only spacecraft to previously visit was the Mariner 10, which completed its mission in 1975, having mapped only 45 per cent of the planet's surface with what was, by today's standards, low-resolution equipment. Messenger is planning to fill in the gaps.

Launched on 3 August 2004, the spacecraft is designed to deal with the very specific problems of travelling to the inner solar system. Probes travelling to the outer planets have to accelerate to high speeds to travel across the vast distances in a reasonable time. They do this by a complex series of gravitational slingshots around the inner planets of the Solar System to speed them up. They must also carry enough fuel to slow them down when they reach their destination or, like Voyager, sweep past on a fleeting visit before vanishing off into interstellar space. Messenger has the opposite problem; because of Mercury's proximity to the Sun, the force of gravity causes Messenger to accelerate faster and faster as it approaches Mercury. It is rather like dropping a ball down a deep well. But unlike Mariner 10, Messenger is designed to go into orbit around the scorched planet. This requires Messenger to slow down enough to be captured by Mercury's weak gravitational field – an immensely tall order. It's like having to stop the falling ball halfway down the well.

To achieve this feat of spacecraft navigation, Messenger has been on an extended and complex journey, flying by Mercury three times over the first five years of its mission, each time using Mercury's gravity and orbital speed to slow down enough to enter orbit around Mercury in March 2011. Once in place, the probe is tasked with answering many of Mercury's mysteries; from the suspicion that there may be ice on the poles of this sun-drenched planet, to the theory that Mercury may be shrinking in size. Perhaps most important of all will be the completion of the work started by the Mariner probe in mapping the whole surface of Mercury and filling in the gaps that exist in our understanding of its geological history. As we capture more and more new images from the surface of the planet, it is becoming abundantly clear that one kind of geological feature dominates the surface.

For the last 4.6 billion years, Mercury has been bombarded with countless asteroids and comets. Unlike our planet's protective blanket, Mercury had nothing to shield it from the onslaught. When a meteorite hits naked Mercury, there is no atmosphere to break it up or slow it down; it strikes the ground at full speed and intact. As Messenger's images are revealing in beautiful vivid detail, the whole history of the planet's violent past is laid out on its surface, a world pitted with hundreds of thousands of craters inside craters, inside craters.

Mercury was damned from the start. It's too small and too hot to have retained any meaningful traces of atmosphere. Earth, though, is big and cold enough to have retained this envelope of gases that allows living things to evolve and to use that atmosphere to breathe and to live.

That's not where our luck ends, though. There is a place out there in the Solar System whose atmosphere began with the same ingredients as our own. A planet roughly the same size as Earth, and not much nearer the Sun. On this planet the characteristics of its atmosphere and place in the Solar System have been only slightly remixed, and yet it is a world that couldn't be more different from our own ◉

DEFENCELESS PLANET
Mercury is constantly being hit by comets and asteroids. It suffers these onslaughts because it has no atmosphere, which heats up the asteroid and breaks it into smaller pieces.

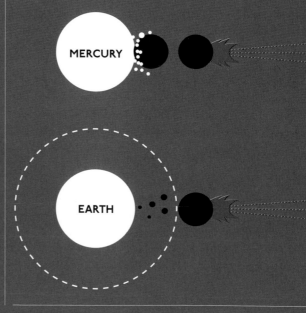

132

BELOW: The first visitor to Mercury was the Mariner Venus/Mercury spacecraft, also known as Mariner 10. It was launched on 3 November 1973 from NASA's Kennedy Space Center, and three months later it flew by Venus, managing to map only 45 per cent of the planet's surface with what modern technology now views as low-resolution equipment.

BELOW AND BOTTOM: The Messenger space probe has relayed spectacular and informative images of Mercury in its three fly pasts so far. The image below shows the Brahms Crater, with a diameter of ninety-eight kilometres (sixty miles).

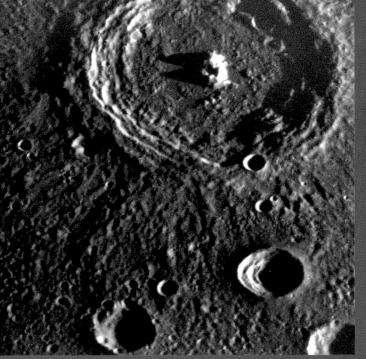

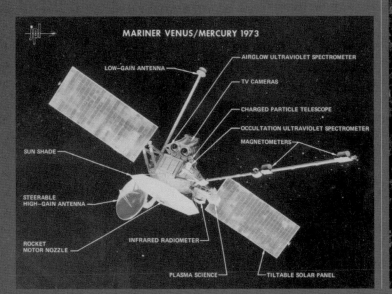

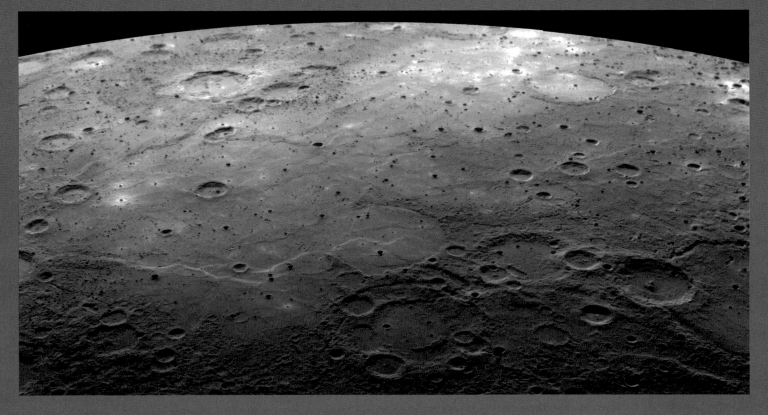

GOLDILOCKS ATMOSPHERES

Roughly 108 million kilometres (67 million miles) from the Sun sits Venus, the brightest planet in our night sky. Completing its orbit every 225 days, Venus is luminous enough to cast shadows on Earth as it reaches its maximum brightness just before sunrise or just after sunset. Venus and Earth share many similarities. We sit next to each other in space, we were formed from the same material, we are roughly the same size and we also share a similar mass and therefore gravitational field. But that's where any similarities end.

Venus is a tortured world where thick clouds of sulphuric acid are driven by high-speed winds and temperatures at the surface are hot enough to melt lead. It's not surprising that Venus is often known as Earth's evil twin. The reason for this hellish difference between these two superficially similar worlds is primarily down Venus's atmosphere, which evolved along a very different path to our own.

On 10 August 1990, the Magellan space probe began a four-year mission in orbit around Venus. Its aim was to give us the first images from beneath the shroud of cloud that had hidden our view of Venus for centuries. The images that Magellan sent back revealed a tortured landscape of volcanoes and impact craters. Beneath the clouds of sulphuric acid the landscape bore little resemblance to anything we can see here on Earth.

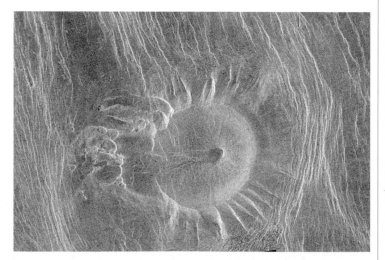

ABOVE: This picture, taken by the Magellan space probe on its mission in the early 1990s, shows our neighbouring planet Venus. During its four-year voyage, Magellan captured images of the Face of Venus, which clearly show the presence of volcanoes. Here is an example of a fairly common type of Venusian volcanic feature, known as a 'tick'. It is a volcano that is probably about thirty kilometres (twenty miles) wide at its peak, encircled by ridges and valleys that radiate down its sides and give it the insect-like appearance from which it gets it name.

Despite almost fifty years of interplanetary travel, sending images like this across forty-two million kilometres (twenty-six million miles) of space still remains an extraordinary feat of engineering. Venus was the destination for the first interplanetary mission when Mariner 2 reached the planet in December 1962. The hardware on Magellan may have been light years ahead of Mariner's (the very best that the late twentieth century could offer), but the mathematics needed to carry these detailed images back to Earth was the same and much, much older.

Almost every image that NASA has ever sent across space has relied on the work of the French mathematician and physicist Joseph Fourier in the early nineteenth century. The Fourier transform is a beautiful piece of mathematics

Beneath the clouds of sulphuric acid the landscape bore little resemblance to anything we can see here on Earth.

that was originally developed with no practical application in mind, and yet today Fourier's work can be found in virtually every electronic image we see; from our family photographs stored as JPEGs to the images we gather from the far reaches of the Solar System. It is the mathematics that drives our ability to compress huge amounts of information into files that are small enough to send around the world, or even around the Solar System.

However, when the Magellan space probe went to work, Fourier's contribution to our understanding of Venus wasn't just limited to the technology required to send images flooding back to Earth. In 1824, Fourier became the first scientist to describe an effect that is both crucial to our understanding of Venus and vital for the future health of our own planet.

The greenhouse effect has become a well-known contemporary phrase, one that is now synonymous with global warming; but in fact the idea originally came from the notebooks of Fourier in the nineteenth century. He was the first scientist to suggest that gases in the Earth's atmosphere might cause the planet to heat up. In doing so, Fourier paved the way not only towards an understanding of our climate, but also the extreme effects of the greenhouse effect on our sister planet Venus ◉

BELOW: This computer-simulated image shows the northern hemisphere of Venus. The planet is known as the evil twin of Earth; they are similar in size but Venus is closer to the Sun, which means temperatures on the planet soar to above 370 degrees Celcius. Earth is known as the Goldilocks planet because, unlike its neighbours Venus and Mars, which both experience extreme temperatures, the temperature on our planet is 'just right' for life.

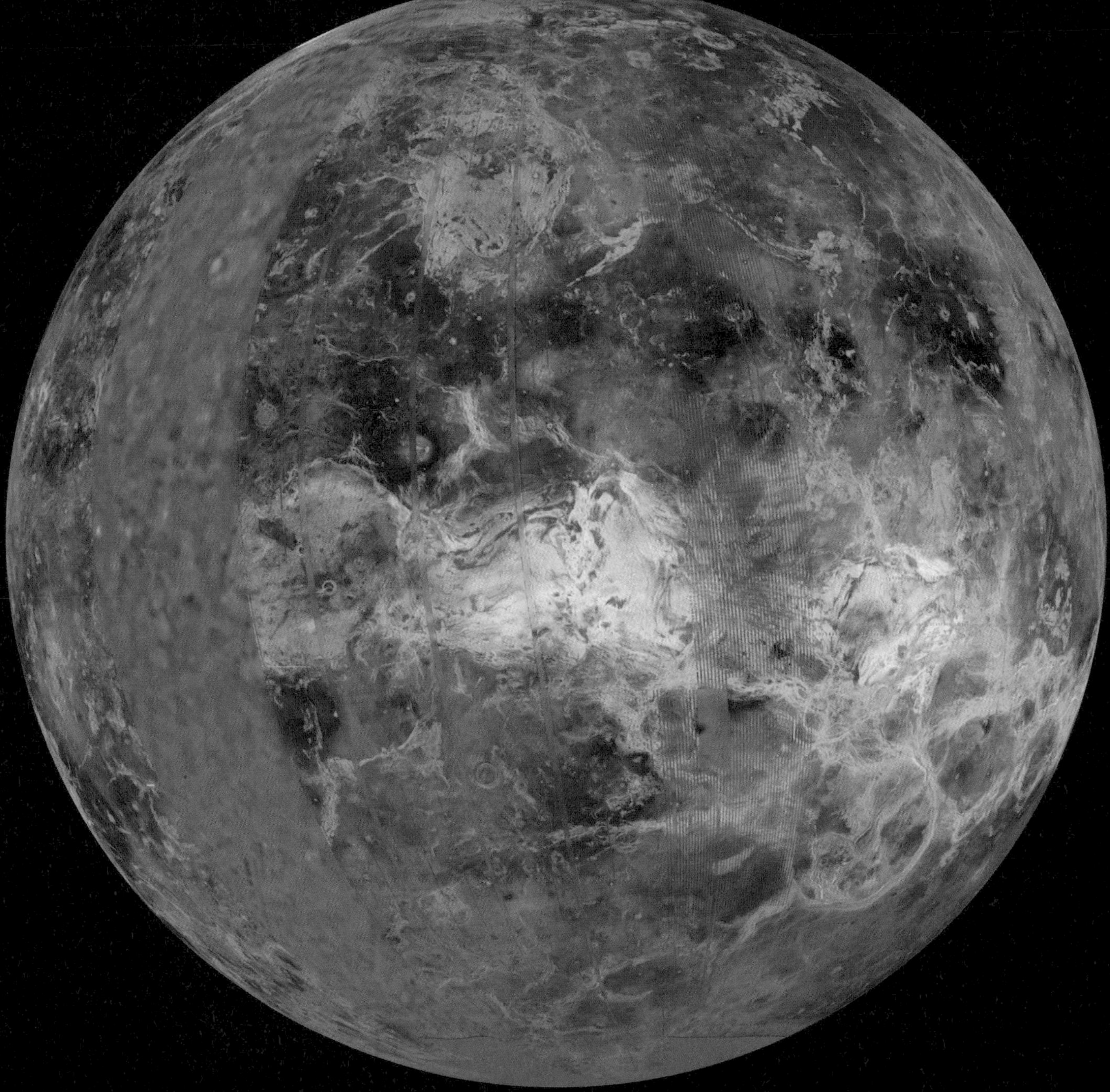

THE GREENHOUSE EFFECT

The greenhouse effect is basically a very simple piece of physics. The gases in a planetary atmosphere absorb the light of some wavelengths and allow others to pass through to the ground unimpeded. Earth's atmosphere is mostly transparent to visible light, which is obvious because we can see the Sun in the sky! Fortunately for life on Earth, much of the damaging, shorter wavelength, UV light is absorbed in the upper atmosphere by ozone. The sunlight allowed through by our atmosphere warms the Earth's surface, which then re-radiates this energy as infrared radiation. Infrared light has a longer wavelength than visible and UV light, and atmospheric gases such as carbon dioxide and water vapour are very good at absorbing it. In other words, some of the heat radiation from the ground is prevented from escaping back into space by so-called greenhouse gases. This means that the atmosphere gradually heats up, raising the temperature of our planet.

On Earth, the greenhouse effect is essential to our survival. Without the concentrations of greenhouse gases in our atmosphere we have today, our planet would be on average thirty degrees Celsius colder; far too cold to support life as we know it. A little greenhouse effect is a good thing, but if the concentrations of gases such as carbon dioxide are raised too much, we only have to look to our nearest neighbour to see the devastating consequences.

Venus' atmosphere is flooded with greenhouse gases. The rising temperatures would have long ago boiled away its oceans, pumping water vapour into the atmosphere. Carbon dioxide from thousands of erupting volcanoes added to the stifling mix. Venus grew hotter and hotter – far hotter than its position closer to the Sun than Earth would merit. The planet slowly choked to death ◉

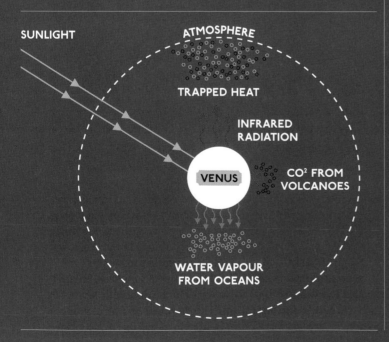

SUNLIGHT

ATMOSPHERE

TRAPPED HEAT

INFRARED
RADIATION

VENUS

CO_2 FROM
VOLCANOES

WATER VAPOUR
FROM OCEANS

BELOW: On Earth, greenhouse gases are essential; without them our planet would be thirty degrees Celsius colder, which would make life unsupportable. Venus, however, suffered at the hands of excess greenhouse gases, choking to death on them.

THE RED PLANET

LEFT: On the rover's 2,052nd
Martian day, here is a view of NASA's
Mars Exploration Rover Spirit's
robotic arm.

The Namib desert in Namibia is not the hottest desert in the world, nor is it the driest, but its ancient sand dunes form part of the oldest desert anywhere on the face of the Earth. Stretching over 1,900 kilometres (1,200 miles) along the south-western Atlantic coast of Africa, this vast expanse has been starved of rain for over fifty-five million years. It is a spectacular landscape, made even more fascinating for planetary astronomers because it is an extraordinarily accurate analogue for the surface of Mars. If you want to experience the sensation of standing on Mars here on Earth, at least visually, there is no better location to place your feet than in the crescent-shaped Barchan dunes of the Namib. Your eyes will be transported fifty-five million kilometres (thirty-four million miles) across space to the surface of another world.

ABOVE: In January 2005, NASA's Mars Exploration Rover Opportunity found an iron meteorite on Mars, the first meteorite of any type ever discovered on another planet.

The Rover got close enough to the pitted meteorite, which is the size of a tennis ball and now called 'Heat Shield Rock', to determine that it is mostly made of iron and nickel.

The reason we understand the landscape of Mars in such beautiful detail is because we have first-hand evidence. The first successful landing on Mars was made, very briefly, by the Russian Mars 3 probe in 1971, but it only managed to transmit data back to Earth for fifteen seconds. The first truly important and revelatory landings were made by NASA's Viking probes in 1976. Viking searched unsuccessfully for signs of life on the red planet, although some of its results remain controversial to this day, and there are those who believe that Viking may have seen such evidence for life. Until the next biology-focused missions land on the Martian surface, we will probably never know.

In January 2004, the Opportunity Rover and the Spirit Rover touched down on the red planet and began the most intense and direct exploration of any landscape other than our own. Designed to roam Mars for ninety days, these little rovers are undoubtedly two of the most successful spacecraft ever launched. At the time of writing this book, in summer 2010, both rovers are still in contact with Earth, although Spirit appears to be stuck in the sands of Mars. For spacecraft that were not expected to see out 2004, this is nothing short of astonishing and one of the great feats of human exploration. Day after day, year after year, Spirit and Opportunity have driven across the surface of Mars and sent back exquisitely detailed images. Again and again the images reveal landscapes that have an eerie familiarity. Mars has vast sand dunes, enormous volcanoes, giant ice sheets, canyons and river valleys. It is a dry, frozen version of our home, covered in red dust and sand, entirely familiar and yet entirely inhospitable to human life. The barren landscape is due to the fact that today Mars has virtually no atmosphere. Yet as the rovers have searched the planet they have found compelling evidence to suggest that Mars hasn't always been this way.

In January 2005, the rover found a rock that turned out to be a nickel iron meteorite; four years later, in August 2009, it found another, which was estimated to be ten times bigger than the first. This makes it the biggest meteorite ever discovered on a planet other than our own, and its very existence makes no sense given what we know of the Martian atmosphere today.

Mars' atmosphere is incredibly thin. It consists of 95 per cent carbon dioxide, 3 per cent nitrogen and 1.6 per cent argon, with only traces of water and oxygen. In comparison to Earth, the mass of the atmosphere is tiny; 25 million million tonnes compared to the Earth's 5,000 million million tonnes.

BELOW: This picture is one of the first captured by the camera attached to the Pathfinder lander not long after it touched down on Mars on 4 July 1997. In the foreground you can see multiple images of the same little space rover, named Sojourner. On the horizon are two hills beyond the dusty rocky landscape of the planet's surface.

If you want to experience the sensation of standing on Mars here on Earth ... there is no better location to place your feet than in the crescent-shaped barchan dunes of the Namib.

THE THIN BLUE LINE

Standing on Mars you would be exposed to less than 1 per cent of the Earth's surface atmospheric pressure – equal to the pressure we'd experience at an altitude of thirty-five kilometres (twenty-two miles). If the meteorite found by Opportunity had hit the planet today there would have been nothing to slow it down. The Martian atmosphere is too thin to provide any significant breaking force to a meteorite of this size, and so it would have been travelling so fast when it hit the surface that it would have disintegrated on impact. It simply shouldn't be there!

The most likely explanation for this seeming paradox is that at some point in the past when this meteorite hit Mars, the atmosphere was significantly denser; dense enough to slow it down so that it could land on the surface intact.

If this is correct, then why did Mars lose its thick atmosphere and become the barren planet we see today? Atmospheres are fragile ghosts around a planet and there are many ways for them to be disrupted and lost to outer space. When you realise how fragile they are it begins to feel like a miracle that we've still got ours at all. It is thought that the reason Mars lost its atmosphere is down to the red planet's interaction with the powerful and far-reaching influence of our sun.

The solar wind is a stream of super-heated, electrically charged particles that constantly flow away from the Sun at over a million kilometres per hour. This wave of smashed atoms may be invisible and appear innocuous to us here on Earth, but it has the power to strip a planet of its atmosphere. We are protected by an invisible shield that completely surrounds our planet, known as the Earth's magnetic field. The origin of the Earth's magnetic field is in its molten iron core. This magnetic shield is strong enough to deflect most of the solar wind that comes our way (see page 52), in stark contrast to Mars ◉

HOW MARS LOST ITS ATMOSPHERE

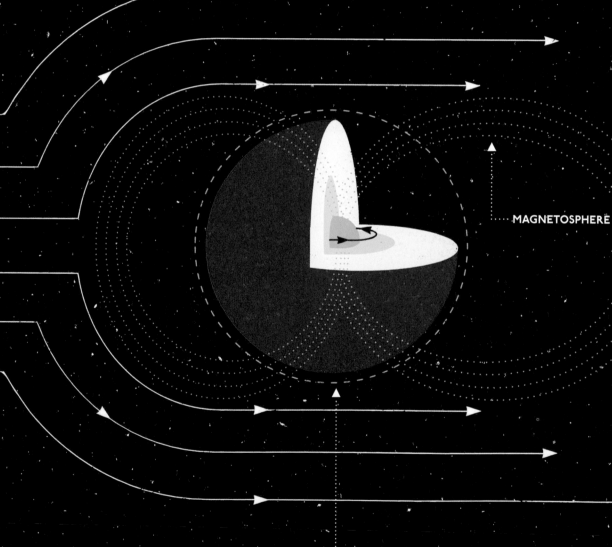

SOLAR WINDS

MAGNETOSPHERE

ATMOSPHERE

MARS BEFORE

About four billion years ago, Mars had a molten core. Mars, after all, was formed by the same processes as Earth, from the same material around the same star and its molten core would have generated a protective magnetic field. There is, however, one crucial difference between the planets; Earth is nine times bigger, with Mars' total surface area being the size of the area of the dry land on Earth. This size difference is crucial because the larger the ratio of the surface area to the volume of an object, the quicker it will lose its heat.

Early in Mars' history the planet lost its internal heat through its surface and out into space, its core solidified, electric currents could no longer flow and its magnetic field vanished. Without these defences, the solar wind would have blasted Mars and stripped its atmosphere, and with no atmosphere to insulate it, this once Earth-like world was transformed into the frozen desert we see today, a shadow of its former self. Although Mars has lost most of its atmosphere, those few molecules of atmosphere that remain still have the power to sculpt its surface.

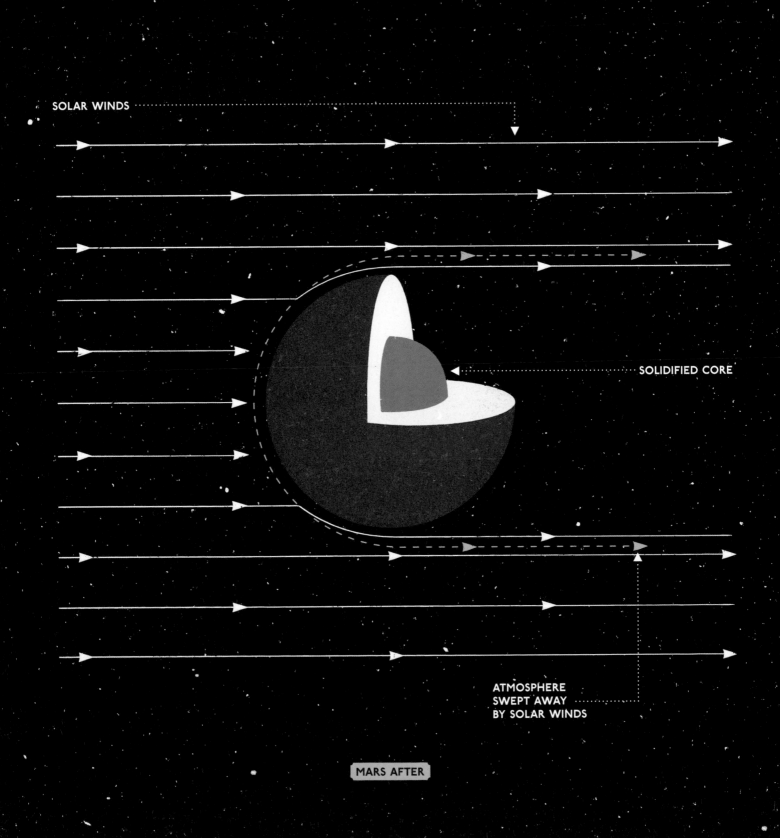

SOLAR WINDS

SOLIDIFIED CORE

ATMOSPHERE
SWEPT AWAY
BY SOLAR WINDS

MARS AFTER

THE STORMY SOLAR SYSTEM

Weather is a feature of every planet with an atmosphere – however tenuous and diffuse it might be. Wind, storms, clouds and even rain can be found on worlds beyond our own. Wherever there is an atmosphere there is a delicate and complex interaction between the heat of the Sun, the surface of the planet and the swirling mass of gas that surrounds it. It is easy to see and understand how our world is transformed as the huge mass of air moves across the Earth's surface. But as we look out into the Solar System, we have discovered that it only takes the slightest trace of an atmosphere to produce extraordinary weather.

RIGHT: A deadly windstorm strikes central Kansas, USA.

····· JUPITER ····· EARTH

Jupiter's atmosphere is mostly made up of molecular hydrogen and helium and is many thousands of kilometres thick. It is divided into four layers (subject to altitude) and is in a constant state of seething motion and experiences cyclones, anticyclones, storms and lightning.

THE GREAT RED SPOT
Large enough to swallow our planet three times over and to be visible from telescopes on Earth, this is an anticyclonic storm that has persisted for at least 180 years, and possibly as many as 350.

146

THE THIN BLUE LINE

BELOW: The largest planet in our solar system, Jupiter is a giant ball of gas and liquid. Its surface is made up of dense red, yellow, brown and white clouds and boils with storms that are far more powerful than anything we experience on Earth.

JUPITER: A PLANET OF WEATHER

Weather is a feature of every planet with an atmosphere, however tenuous and diffuse it might be. Wherever there is an atmosphere there is a delicate and complex interaction between the heat of the Sun, the surface of the planet and the swirling mass of gas that surrounds it. To experience the most extreme and violent weather in the Solar System, we need to visit Jupiter, the largest planet around our star. This giant is over 140 thousand kilometres (87 thousand miles) in diameter, dwarfing the Earth in volume. Over thirteen hundred Earths could comfortably fit inside Jupiter. Primarily made of hydrogen and helium, Jupiter is almost all atmosphere. There is no thin blue line; instead the Jovian atmosphere is many thousands of kilometres thick and in a constant state of seething motion, boiling with gigantic storms.

Yet this most alien world shares a feature with our own planet. Jupiter crackles to the sound of electrical storms. Bolts of lightning thousands of times brighter than lightning here on Earth illuminate the Jovian sky. These gigantic storms may look alien through the eyepiece of a telescope, but the forces that drive them are identical to the forces that drive storms here on Earth.

If there is warm moist air deep in an atmosphere it will start to rise, and as it rises it cools and the moisture condenses out to form clouds. That rising air leaves a gap beneath it, a low-pressure area, and so more warm moist air is sucked in, which this fuels the beginnings of a storm.

Here on Earth, the storm systems are driven by the power of the Sun. It is the Sun heating the Earth that creates the convection currents that churn our atmosphere into action. Without the Sun's energy, our planet would be a much calmer place. Jupiter, by contrast, is five times further away from the Sun, which means it receives twenty-five times less solar energy per square metre. This means that you might expect its storms to be considerably weaker. Intriguingly, we have discovered that the opposite is true; the storm systems on Jupiter are far more powerful than anything we have experienced on Earth. So what mechanism could possibly power these intensely violent storms?

The secret to Jupiter's storm-tossed atmosphere lies hidden deep within the gas giant. On Earth we have clear boundaries between the gaseous sky, the liquid oceans and the solid ground. On Jupiter there are no such boundaries; Jupiter is a giant ball of hydrogen and helium – a planet built almost entirely of atmosphere. But as you go deep down into Jupiter's atmosphere, something very strange and interesting happens to those gases.

Jupiter's atmosphere is so dense that 20,000 kilometres (12,000 miles) beneath the cloud tops the pressure is two million times greater than the surface pressure on Earth. Under these immense pressures the hydrogen gas in the atmosphere is transformed into a strange metallic liquid. When gases turn into liquids on this colossal scale, vast amounts of energy are released. Think of it this way: you have to put energy into a pan of liquid for it to boil and turn into steam. So if you do the opposite and condense steam back into liquid water, energy must be released. The same is true for gaseous and liquid hydrogen. It is this energy source that creates the convection currents that fuel some of the biggest storms in the Solar System.

The biggest of all the Jovian storms raging at the moment is the Great Red Spot, a gigantic storm 40,000 kilometres (25,000 miles) from east to west and 14,000 kilometres (9,000 miles) from north to south. This giant anticyclone has been raging for hundreds of years and is three times larger than the Earth. It is thought that the wind speeds reach over 400 kilometres (250 miles) per hour as this violent atmospheric feature circles the great planet every ten hours. We do not know why the storm has raged for so long or what makes it so red. It is thought that complex organic molecules formed from methane in Jupiter's upper atmosphere as it reacts with the Sun's ultraviolet radiation could be one factor behind its vivid colour.

The Great Red Spot is an extraordinary example of the violent alien weather that exists around the Solar System, but if we want to experience the most Earth-like atmosphere outside of our planet, then we need to look at a much smaller world. Orbiting the gas giant Saturn, one and a half billion kilometres from Earth, there is a magical frozen world that until recently hid its wonderful secrets beneath a thick impenetrable veil of cloud ◉

TITAN: THE MYSTERY MOON

LEFT: The mysterious moon, Titan, is captured in this false-colour composite created from images taken by the Cassini spacecraft on 16 April 2005. The green areas are where Cassini could see down to the surface; red are areas high in Titan's stratosphere where atmospheric methane is absorbing sunlight; and the blue illustrates the thick, Earth-like atmosphere that surrounds it.

Of all the 170 known moons in the Solar System, we have only been aware of one for any length of time. Our moon has been a source of wonder for millennia; the Moon is the fifth biggest natural satellite in the Solar System, and it dominates our night sky because it is by far the biggest moon in the Solar System in relation to the size of its parent planet. Our moon has a profound effect on the life of our planet, driving the ocean tides that are intimately linked to the cycles of Nature. It may even have been the case that tidal pools were the cradle for the origin of life on our planet.

It is also the only world beyond Earth on which humans have stood, and we discovered a dead, dusty world. Buzz Aldrin's famous description of the Lunar surface as 'magnificent desolation' is a savage yet romantic and perfectly apt description of the Moon's harsh beauty. Our only satellite is riddled with craters, the ancient remains of volcanoes and an atmosphere so tenuous that it is virtually indistinguishable from a vacuum. The very word 'moon' conjures a sense of a lifeless, inert world, a world as different to our dynamic planet as could be.

We have long thought of our moon as the archetypal moon of the Solar System, perhaps because we know it so well. This has meant that when we began our journeys into space many scientists expected the planets to be the stars of the show. The majority of those moons out there were thought to be dead, uninteresting worlds, but this could not be further from the truth. As is always the case with

exploration, you genuinely don't know what you'll find until you actually go there. This is as true for the frozen outer solar system as it is for the most distant and isolated places on Earth. However, as we have begun to visit these worlds and fly spacecraft to within a few kilometres of their surfaces, we've found that the moons are an astonishingly interesting, varied and fascinating bunch of worlds.

One such place is Titan – Saturn's largest moon and the second largest moon in the Solar System. Bigger than the planet Mercury, this giant moon remained a virtual mystery until the Cassini spacecraft and its tiny sister, Huygens, arrived in the Saturnian system in 2004. The picture opposite, taken by Cassini in April 2005, illustrates why this world has always remained an intriguing but mysterious place, unique amongst the moons of the Solar System.

Surrounding Titan is an atmosphere that is 600 kilometres (370 miles) deep and four times as dense as that of the Earth. Titan is a magical place – a moon circling around a distant planet in the outer reaches of the Solar System, with an atmosphere that is more substantial than our own. It is the most Earth-like atmosphere we know of anywhere out there in space; a thick blue line rich in nitrogen and containing methane.

It seems almost beyond imagination that a world this small should be able to hold on to such a dense atmosphere. Mercury is too small and hot to have the gravitational grip needed to hold on to its atmosphere and Titan, although larger in volume, is half the mass of Mercury and so has an

BELOW: As the sunlight glows and scatters through the periphery of Titan's atmosphere, it produces a circle of light around the planet.

Chinese lanterns explain a simple bit of physics; high temperatures mean fast-moving molecules. By lighting the fuel beneath the lantern, the air inside heats up, making the molecules inside whizz around, increasing the pressure in the lantern.

even weaker grasp on the atoms of gas surrounding it. The reason that Titan has its wonderful atmosphere is because it lies in a much colder region of the Solar System – and that makes all the difference.

HOW TITAN KEEPS ITS ATMOSPHERE

The temperature of a gas is essentially a measure of how fast molecules are moving around – the higher the temperature, the faster they are moving. The speed of the molecules is also related to the pressure the gas exerts, because pressure is simply the effect of the molecules smashing against something, and the faster they are moving, the harder they smash into things and the higher the pressure. School chemisty lessons will have left you with an equation imprinted in your brain that summarises all of this – it's called the ideal gas law. It says that the pressure times the volume is proportional to the temperature – $PV = nRT$ – where P is the pressure, V is the volume, n is the number of molecules (measured in an obscure quantity called 'moles'), T is the temperature and R is a number known as the ideal

gas constant. What this says in words is that if you keep the volume of a container fixed and raise the temperature, you have to raise the pressure or lower the number of molecules to keep everything in balance.

You can demonstrate what all this means beautifully with a Chinese lantern. If you light the fuel beneath the lantern, the air inside the lantern heats up. This means that the molecules inside start whizzing around faster and the pressure inside the lantern begins to increase. But the lantern is open at the bottom, and therefore the pressure inside the lantern must remain the same as the pressure outside. The pressures equalise by molecules of air rushing out of the bottom of the lantern and disappearing off into the atmosphere. If you look at our equation, n is decreasing in order to allow T to increase and keep everything else the same. Because molecules are constantly rushing out of the lantern, it weighs less and less as time goes by and eventually it is light enough to float gently off into the sky. Thus Chinese lanterns explain a simple bit of physics; high temperatures mean fast-moving molecules.

BELOW: Chinese lanterns are a beautiful example of the ideal gas law in action.

One and a half billion kilometres away from the Sun, Titan barely feels the warm glow of our star. Out here, the Sun is barely more than another star in the sky, making Titan a very cold place indeed. This means its atmospheric molecules are moving very slowly relative to ours. If Titan were in the same region of the Solar System as we are it would not be able to hold on to its atmosphere. It is a much less massive body than Earth, and so it has much a weaker gravitational pull. If the Sun heated Titan to Earth-like temperatures its atmosphere would soon vanish into space because it could not hang on to the fast-moving molecules. Move fourteen hundred million kilometres further out into the darkness to Titan's current position, however, and the weakness of Titan's gravitational pull is offset by the fact that its atmospheric molecules are moving around much more slowly than ours, allowing Titan to hang on to its dense atmosphere.

It was first suspected Titan had an atmosphere over 100 years ago. The Spanish astronomer Josep Sola noticed a phenomenon known as the 'limb darkening' of Titan in 1903. He suspected that the transition in the intensity of the light he observed from the centre of Titan to its edge indicated that there was a layer of gas surrounding it. It took the work of an extraordinary astronomer with extraordinary eyesight to provide the first direct evidence of Titan's atmosphere, though. Gerard Kuiper was famed for his ability to see things that no one else could see. His incredibly acute eyesight allowed him to see stars with his naked eye that were four times fainter than those visible to virtually anybody else.

As well as having the Kuiper Belt, the region of planetoids and asteroids beyond Neptune, named after him, Kuiper was also the first to gather the spectroscopic data that confirmed the presence of an atmosphere on Titan. He was even able to estimate the pressure on the surface of the Moon. With the thick cloud of gas shrouding Titan, it was difficult to explore the secrets of this intriguing moon any further using telescopes based on Earth. Nobody could see through the clouds. The Voyager space probe made the first detailed observations, but even from nearby, only the cloud tops were visible. It took a mission of even greater audacity to reveal the world beneath the haze ◉

JOURNEY
TO TITAN

BELOW AND BOTTOM: The man behind the mission: Ralph Lorenz was one of the team who designed the Huygens probe that was launched in 1997. The probe sent back unique images of the planet from a distance as it flew through its atmosphere, and close-ups upon landing on its surface.

RIGHT: This set of images was taken by the Huygens probe on 14 January 2005. The pictures show angles north, south, east and west at five different altitudes above Titan's surface, revealing the planet to have lakes and dunes and a landscape not unfamiliar to us on Earth.

In 1997, the Cassini space probe began its journey to Titan. It carried with it the Huygens probe, a lander designed to set down on the surface of the enigmatic frozen moon. Huygens remained dormant aboard Cassini throughout the six-and-a-half-year journey to its release point above Titan, but on Christmas Day 2004, Huygens separated from Cassini and began the bumpy ride through the most intriguing atmosphere in the Solar System. For the next twenty-two days Huygens gently coasted towards the moon, with nothing switched on except for a 'wake-up' timer, programmed to awake Huygens fifteen minutes before it hit the atmosphere.

As the tiny probe approached the end of its billion-kilometre journey, it deployed a parachute to slow its descent and switched on systems designed to have power for just 153 minutes. Huygens gently swayed down, the thick clouds parted and the surface of Titan was revealed for the first time.

Huygens took two and half hours to reach the surface, but aerospace engineers are masters at building machines that exceed their design specifications, so while Huygens should have sent data from the ground for only the remaining precious few minutes, it actually delivered a unique view of our solar system for over an hour and a half. On the opposite page are some of the images taken by Huygens on its descent. The world it revealed was more familiar than we could have imagined.

One of the first scientists to see these incredible images was a man who helped to design the probe, Ralph Lorenz. I was lucky enough to hear first-hand from Ralph his extraordinary account of how it felt to be the explorer of another world as

those first images came through. He told me, 'It was amazing because we … we had no idea what to expect. We didn't know whether it would be cratered like the moon or just a flat expanse of sand, and then these first pictures came back and it was just astonishingly familiar.'

In the Huygens images you can clearly see rounded stones dotting the landscape. They're smooth and look like they have been eroded by tumbling water, similar to pebbles and stones found on river beds on Earth. It seemed for all the world as if Huygens had landed on a river channel, but this interpretation was initially disputed. Surely there could be no rivers on this frozen moon? Science is not a matter of opinion, however, and the evidence was so overwhelming that it was quickly accepted. This was an extraordinary discovery; evidence of flowing rivers had never been found before on a moon, but this wasn't the only surprise Titan held in store ◉

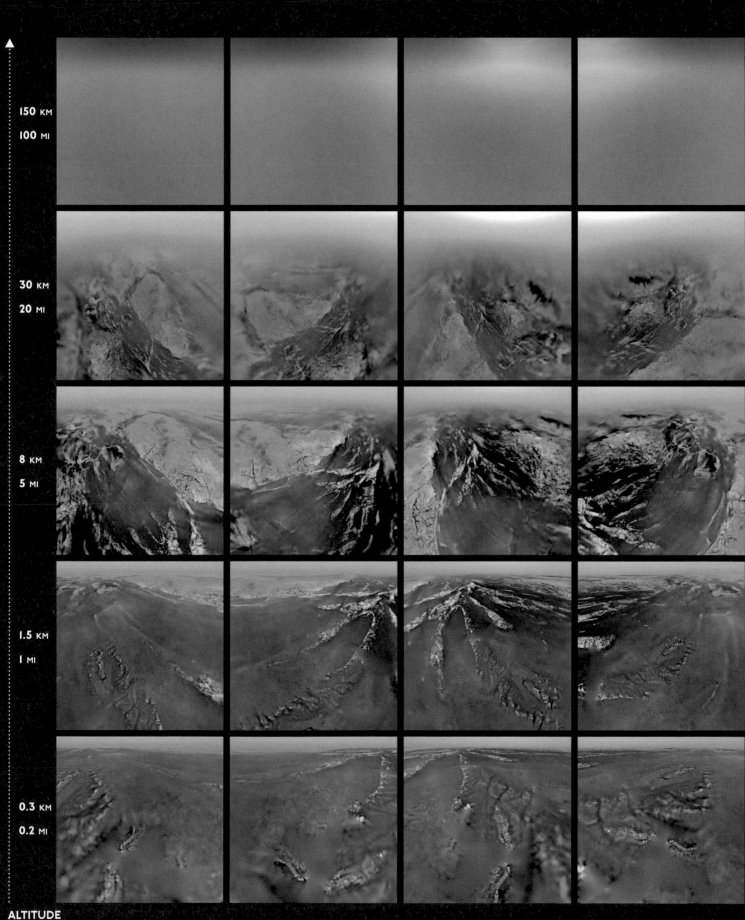

WEST NORTH EAST SOUTH

150 KM
100 MI

30 KM
20 MI

8 KM
5 MI

1.5 KM
1 MI

0.3 KM
0.2 MI

ALTITUDE

THE THIN BLUE LINE

The Matanuska glacier in Alaska is one of the most beautiful places on our planet; a whole landscape that stands testament to the erosive power of ice and rock rolling down a valley over hundreds of thousands of years. The reason that glaciers can exist and sculpt our planet's surface is down to the delicate balance of the Earth's atmosphere.

THE MYSTERY OF TITAN'S LAKES

Our planet is just at the right temperature and pressure to allow water to exist at the surface as a solid, a liquid and as a vapour in the clouds. This happy accident of temperature and pressure allows the Sun to heat up the oceans, raise the water vapour high above the surface as clouds, move it over the tops of the highest mountains and return it to the landlocked ground as rain. The falling rain can then turn to solid ice, become a glacier and sweep down the valley to sculpt astonishing landscapes like the Alaskan Matanuska glacier.

There is a narrow range of temperatures and atmospheric pressures around which substances can exist as solid, liquid and gas simultaneously on the surface of a planet or moon. Because of this delicate balance, worlds that have just the right combination of temperature and pressure to allow water or some other substance to exist as solid, liquid and gas at their

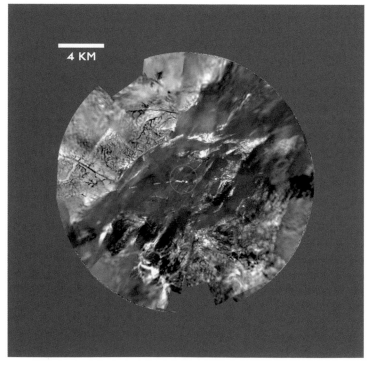

4 KM

surface are extremely rare and precious places. Titan is just such a place – it has the perfect temperature and pressure to allow something to exist that has never been seen before on a world beyond Earth.

The picture above was taken by Cassini in June 2005. The images taken by him at this time have become some of the most important and fascinating in the history of space exploration. The interesting thing on the one above are the black blobs. Immediately, the Cassini scientists were hooked

ABOVE LEFT: The white tracks on this image of Titan's surface show the path of the Huygens probe. The dark, narrow, linear markings on the image have been interpreted as channels cut through the brighter terrain. The complexity of this network of channels across the moon's surface suggests the presence of methane 'rain' and possibly springs.

ABOVE: On Titan the atmospheric pressure of the moon means that methane exists as a solid entity there. So the gigantic lakes on Titan are not filled with water, but liquid methane.

and asked many questions, but an explanation as to what these dark patches were had to wait just over a year until July 2006, when Cassini passed over the same area again and took more images. These radar images showed the north pole of Titan, and the huge black areas are once more visible. In this case the black denotes where radar waves bounced on to Titan's surface from Cassini were not reflected back to the spacecraft, and there is only one really good explanation for that. These features are incredibly flat, so any surface detail at all would cause radar signals to be bounced back. What Cassini saw were surfaces of liquid – the first observation of lakes on the surface of a body other than the Earth in the Solar System (see pages 158–159).

These lakes cannot, of course, be lakes of liquid water, because the surface temperature on Titan is -180 degrees Celsius. At these temperatures, water is frozen as hard as rock. So if these expanses of black in the pictures of Titan's surface are not lakes of water, what are they?

BELOW: Methane is everywhere throughout the Solar System, but on Earth it is an unstable, highly flammable gas. Put a match to methane gas and watch it go up in flames.

RIGHT: These two radar images were recorded by the Cassini radar on 21 July 2006 and are believed to show very strong evidence for hydrocarbon lakes on Titan. The dark patches are scattered all over the high latitudes around Titan's north pole. Some scientists believe that these may have been formed by liquid methane or ethane, particularly near the colder polar regions.

Lake Eyak in Alaska on Prince William Sound is a tranquil place to be in the early morning. The pine-covered foreground hills rising up from the lake occasionally part to reveal higher jagged peaks dappled with snow, even in early autumn, with an Alaskan chill over the waters.

Beyond the picturesque, Lake Eyak is a great place to come to collect a substance that we know is very abundant on Titan. Methane is common throughout the Solar System, and here on Earth it exists as a gas that bubbles up from the depths of Lake Eyak. The floor of the lake is covered in rotting vegetation. The dead leaves, broken trees and twigs are broken down by bacteria, whose metabolic processes produce large quantities of this volatile gas. It's easy to collect the methane that bubbles up – simply tip a boat upside down and leave it overnight in the lake. In the morning, the boat will be filled with methane gas.

It's even easier to show just how unstable methane is here on Earth. Put a match to a bag of methane and in the presence of oxygen you get what chemists call an exothermic reaction. Methane plus oxygen goes to water plus carbon dioxide, and some energy. In other words, it burns.

The Earth's temperature and atmospheric pressure mean that methane can only exist as a highly flammable gas, but on Titan the characteristics of methane are very different indeed. Titan's combination of atmospheric pressure and temperature is perfect to allow methane to exist as a solid, a gas and a liquid. The images Cassini captured are gigantic lakes of liquid methane, the first-ever discovery of a liquid pooling on the surface of another world in the Solar System.

The largest body of liquid on Titan is a lake known as Kraken Mare. At over 400,000 square kilometres (154,400 miles), it is almost five times the size of Lake Superior, North America's greatest lake. It is a true wonder of the Solar System; a vast expanse of liquid on a world almost a billion kilometres from home.

On Titan, methane plays exactly the same role that water does here on Earth. Where we have clouds of water, Titan has clouds of methane with methane rain; where we have lakes and oceans of water, Titan has lakes of liquid methane; and whereas here on Earth the Sun warms the water in the lakes and oceans and fills our atmosphere with water vapour, on Titan the Sun lifts the methane from the lakes and saturates the atmosphere with methane. On Earth we have a hydrological cycle, whereas on Titan there is a methanological cycle.

There is no doubt that rain would be an absolutely magical sight on Titan. The atmosphere is so dense and the gravity of the moon is so weak that the drops of methane rain would grow to about a centimetre in size and would fall to the ground as slowly as snowflakes fall onto the surface of our own planet. Thousands and thousands of gallons of liquid

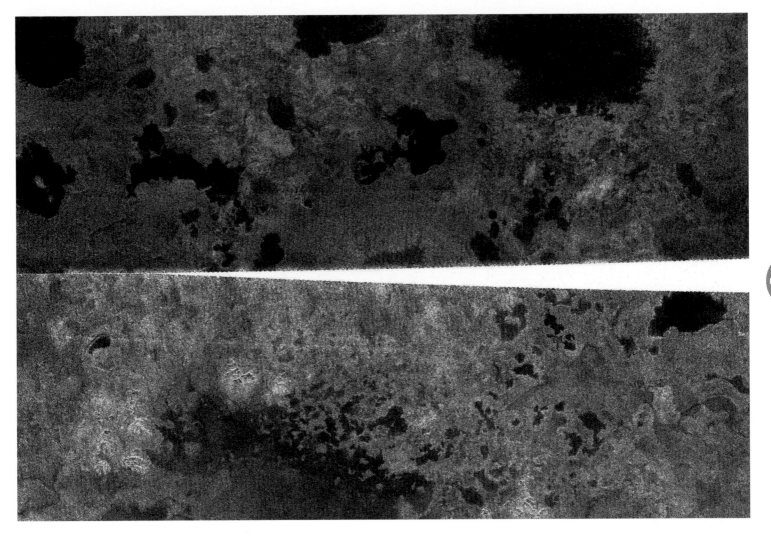

TITAN'S METHANE CYCLE
Titan's methanological cycle follows the same pattern as our hydrological one.

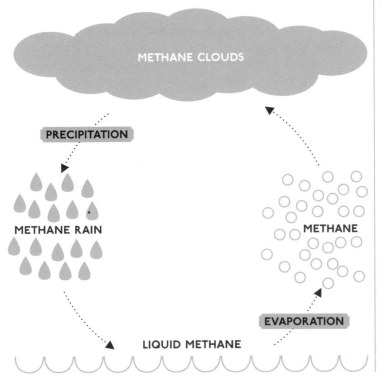

METHANE CLOUDS

PRECIPITATION

METHANE RAIN

METHANE

EVAPORATION

LIQUID METHANE

methane must have slowly rained down on to the surface, making rivers and streams swell and burst, cutting deep gullies into the frozen water landscape.

It all looks and seems so familiar because it is familiar. We are gazing from space on to a landscape sculpted by the same forces of nature and the same cycles that we see here on Earth. The atmosphere of Titan shapes the surface in exactly the same way that our atmosphere shapes the surface of our home planet.

Titan is like a primordial Earth caught in a deep freeze. It's a place with rivers and lakes and clouds and rain. It's a place with water, albeit frozen as hard as steel, and a place of methane, albeit so cold that the methane is a liquid that flows and shapes the landscape just like water does here on Earth. It's almost like looking back in time over four billion years and observing our planet before life began; before our atmosphere was changed by the delicate processes of life into the oxygen-rich canopy of vapours we see today.

Perhaps the most important thing of all about Titan is that we now have two Earth-like worlds in our solar system. One in a warm region, 150 million kilometres away from the Sun, and the other in deep freeze, a billion kilometres away from our star and in orbit around another planet. Surely that must greatly increase the probability that there are other Earth-like planets in orbit around the hundreds of billions of stars out there in the Universe? ◉

CHAPTER 5

DEAD OR
ALIVE

THE HEAT WITHIN

Since the dawn of human history we've been able to gaze up into the night sky, but we're lucky because we're the first generation that's been able to build machines to actually go to those planets and moons. We've found that they're more beautiful, more violent, more magnificent and more fascinating than we could have possibly imagined. The more worlds we study, the more we realise that our solar system is a cosmic laboratory. Even the slightest differences in size or position can create a world radically different from its neighbours.

I n 1540, the Spanish explorer García López de Cárdena was stationed in the tiny outpost of Cibola, in Arizona, when he was asked to conduct a reconnaissance mission. Reports had suggested that there was a large river somewhere north of the camp and so Cardena set off to try and establish the whereabouts of this precious source of water and food. After twenty days of walking northwards, Cardena found what he was looking for. Ahead of him lay the waterway they named the river Tzion. The river, which would one day become known as the Colorado river, was within his sights, but despite days of trying he could not find a way down to the water. The precious waters of the river eluded him by the sheer scale of the drop. Although the mission was a failure, Cardena had become the first European to see one of the greatest wonders on our planet. In his search for water, Cardena found himself standing on the South Rim of the Grand Canyon.

Almost 500 years later, the Grand Canyon has lost none of its ability to awe and inspire. Over five million people a year make the pilgrimage to view this epic landscape and to gaze out across what is undoubtedly one of the most stunning views on Earth.

As well its beauty, a visitor to the Grand Canyon is also looking at an extraordinary example of planetary engineering. Estimates suggest the origins of this valley date back approximately seventeen million years, when the Colorado river began to carve its way through the rock. It is amazing to think that in this short space of time a valley

466 kilometres (277 miles) long, 29 kilometres (18 miles) wide and 1.6 kilometres (1 mile) deep has been etched and carved by nothing more than the action of running water.

Perhaps most remarkable of all the stories the Grand Canyon has to tell is the extraordinary history of our planet etched into the walls of the ravine. From top to bottom, the layers of rock reveal a journey through two billion years of the Earth's history; evidence of rising sea levels and ice ages, ancient swamps and extinct volcanoes, chronicle the ever-changing life of our planet. The walls of the canyon provide us with one of the most complete geological columns on Earth, and despite the wonderful variety of tales it has to tell, one thing connects them all: our planet is alive. It remains today as it has been for billions of years – dynamic, changing and vibrant – a world driven by intense heat at its core and shaped by the journey this heat takes to the surface and beyond.

The great forces that shape our world are universal – all planets and moons share the same basic laws of physics. The level of geological activity, or inactivity, is the very essence of a planet's character. As we've explored the Solar System we have seen how these forces can manifest themselves in many different ways; creating worlds that are more alien than we could ever have imagined and worlds more familiar than we could have guessed. Again and again the wonders of our planet are eclipsed in size and scale by the new wonders we are discovering through our exploration of the Solar System ◉

This is the first image of the largest-known canyon in the Solar System, compiled from a series of pictures taken by NASA's Viking spacecraft in the 1970s. Eight kilometres (five miles) deep, up to 600 kilometres (372 miles) wide and over 3,000 kilometres (1,860 miles) long, on Earth it would run all the way from Los Angeles to New York. The Valles Marineris is a canyon so vast that you could fit our own Grand Canyon into one of its side channels.

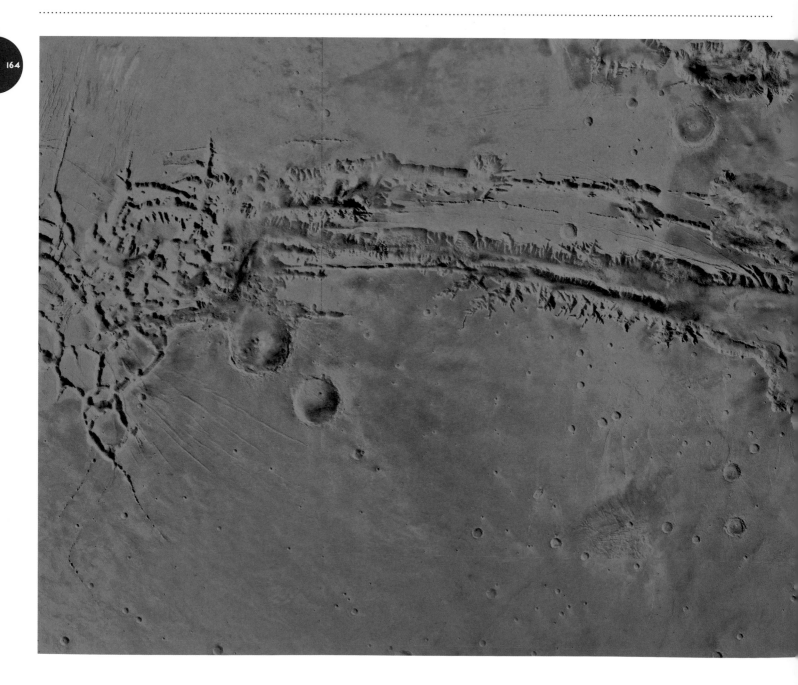

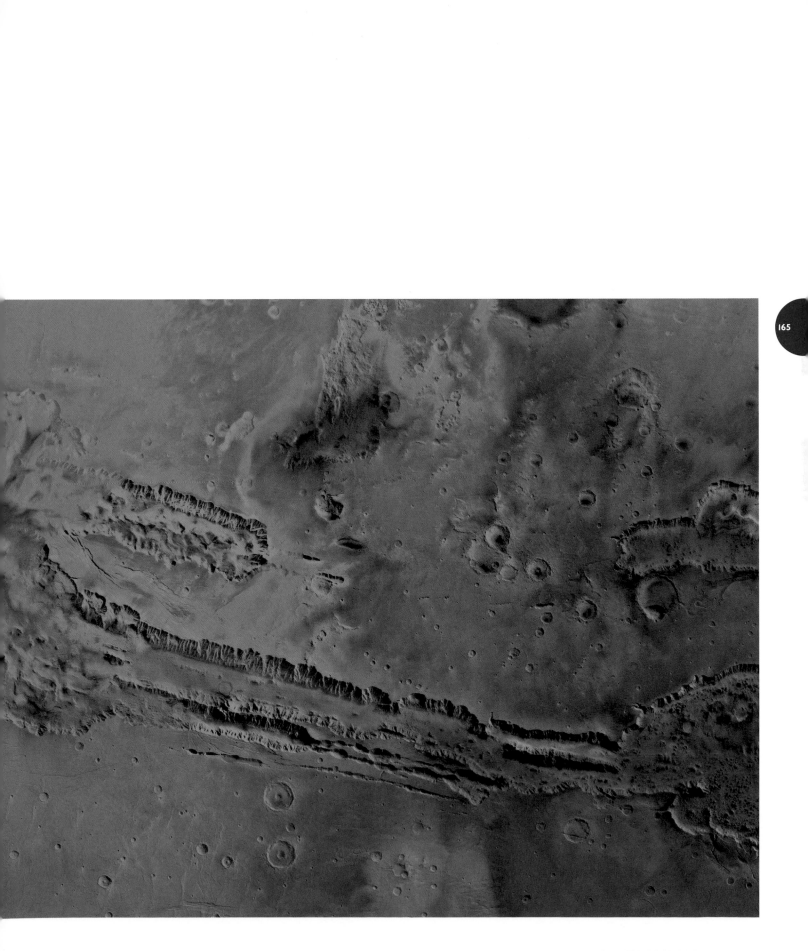

MARS: A FAMILIAR WORLD

The Valles Marinieris is named after the space probe that first discovered it, Mariner 9. Launched in 1971 at the height of the Cold War, Mariner 9 was the first spacecraft to orbit another planet, and the images it sent back revealed the many landscapes and features that we share with Mars. Today, we have three functioning spacecraft orbiting Mars and two rovers, Spirit and Opportunity, on the Martian surface. Together, these are helping us to build a deep and quite profound understanding of the geological evolution of the planet. Unlike any other planet in the Solar System, we have eyes and ears on the surface of Mars and the success of these missions has demonstrated that there really is no substitute for proper exploration on the actual planet. We have sent robotic explorers across millions of miles of space to touch the soil and taste the air. The images they have sent back have allowed us to look up, and gain a new perspective on our solar system.

From the view of our sun setting on the horizon of an alien world to the movement of clouds across the Martian sky, these images reveal the deep similarities between Mars and Earth. These clouds are believed to be composed entirely of water-ice particles, in sizes of several micrometres, formed as part of a band of cloud that occurs near the Equator when Mars is at the coldest, most distant part of its orbit around the Sun. During this cooler part of the Martian year, atmospheric temperatures and the amount of water vapour in the atmosphere allow the formation of large-scale clouds that would not be out of place in the skies of Earth.

Everywhere we look on Mars we are confronted with scenes that remind us of home. It is a landscape that echoes Earth from the very smallest detail to the grandest features, such as the Valles Marinieris. Most scientists now agree that the Valles Marinieris is a tectonic crack that was created in the same way that the plate tectonics here on Earth created the East African Rift Valley. And it is not just tectonic activity that appears to have left its mark on the surface of Mars; we have also found evidence of landscapes formed by water courses and the permanent polar ice caps that ebb and flow with the seasons.

Despite all the similarities between Mars and Earth, it is the differences between these two planets that are most telling. Mars is a cold planet, with an average temperature of minus sixty-five degrees Celsius. It is also a planet with a tenuous atmosphere, compared to Earth.

Mars is now a desolate and dead wasteland, a world where the processes that sculpted its now familiar landscapes seized up long ago. There are no waters to flow, no active volcanoes to erupt, and we have so far found no evidence of life either on the planet's surface or beneath it. Mars appears to be a dead world, a sobering example of how the laws of Nature can play out across our solar system in radically different ways.

BELOW: Mars is currently orbited by three active satellites: Odyssey and Reconnaissance are in normal orbits and Express is in a very elliptical orbit, travelling up to 6,200 miles (10,000 kilometres) away from the surface of Mars.

BOTTOM: Even the skies on Mars are reminiscent of those on Earth, as atmospheric temperatures and water vapour combine to create clouds.

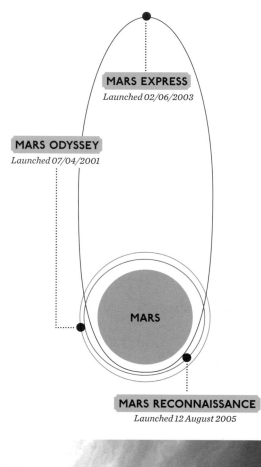

MARS EXPRESS
Launched 02/06/2003

MARS ODYSSEY
Launched 07/04/2001

MARS

MARS RECONNAISSANCE
Launched 12 August 2005

BELOW: This picture of the surface of Mars looks like it could have been taken at Arizona's Grand Canyon; except there we can see the river below that carved this formation, while on Mars we have no explanation for this dramatic rocky landscape.

BOTTOM: The origin of the vast canyon on Mars is unclear, but some scientists believe it began as a tectonic crack billions of years ago.

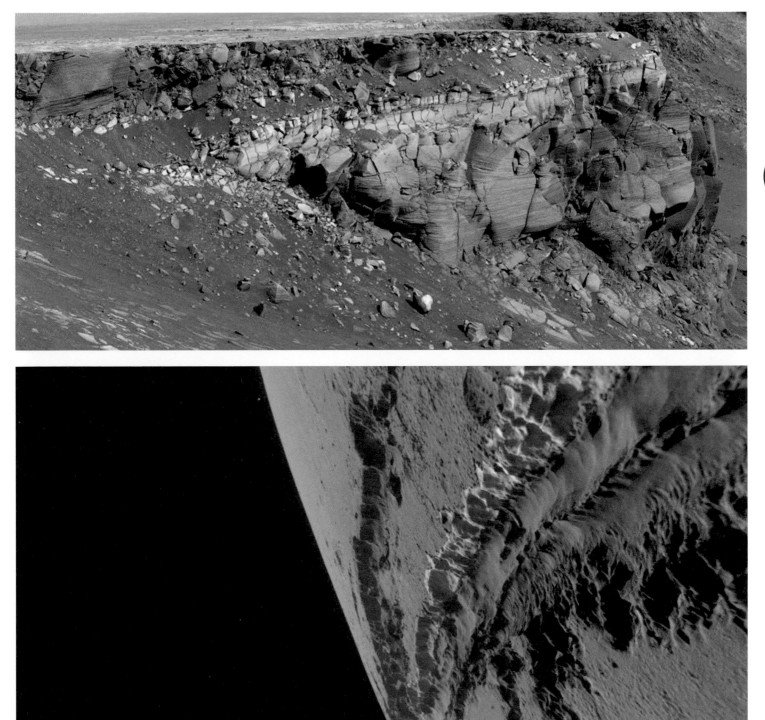

BELOW AND RIGHT: The molten lava spewing out of the volcano Kilauea on Hawaii's Big Island may look destructive, but such eruptions are part of Earth's geological heartbeat. Here is the perfect demonstration of how a planet can be kept alive by nothing more than a flow of heat.

To understand the forces that keep a planet alive, there is no better place on Earth to study than the Big Island of Hawaii. Set amongst the most remote group of islands on the planet, Big Island is part of an undersea mountain range that breaks through the Pacific Ocean to create a chain that runs for over 2,400 kilometres (1,500 miles). This is the perfect place in which to witness how a planet survives by nothing more than the simple flow of heat, because everything on this magnificent island range is created by the intense heat that sits within the Earth's core.

Although the Hawaiian Islands are far from any tectonic plate boundary, they are above a geological hotspot, a narrow stream of lava that is thought to link it all the way to the boundary between the Earth's mantle and its core. As the Pacific Plate drifts over this hotspot, it creates some of the most intense volcanic activity in the world; the magma pushes through, it builds up over time to create the island volcano. As the plate drift moves the volcano away from the hotspot, the magma source is removed and eruptions cease.

The Big Island is built from five shield volcanoes, and by far the most active of these is Kilauea, which means 'spewing'. It's been erupting almost continuously since 1983 and is thought to be one of the most active volcanoes in the world.

Every day you can see molten rock flowing down the side of the mountain, destroying everything in its path. Forests are turned to ash and the Pacific Ocean boils as the lava hits the water and explodes. This might look like widespread destruction, but volcanic eruptions are Earth's geological heartbeat. The surface of our planet has been created and shaped by active volcanoes that make our planet a vibrant living world.

A few kilometres north of Kilauea you can see just what volcanic action can produce, given enough time. Manau Kea lies dormant today, but its scale is testament to the enormous power that exists within the Earth. Although this mountain is four kilometres (two miles) above the surface of the Pacific, it's ten kilometres (six miles) above the surface of the Pacific floor, making it the highest mountain on Earth, but tiny compared to the biggest volcano in the Solar System ◉

THE VOLCANOES OF MARS

Located near the Martian equator is a region known as Tharsis. This vast volcanic plateau, found at the western end of the Valles Marineris, is home to some of the biggest volcanoes in the Solar System, but one volcano dwarfs them all. Its vast outpouring of lava stretches over 600 kilometres (370 miles) wide, but it is the height of Olympus Mons that is truly breathtaking.

It soars twenty-five kilometres (sixteen miles) into the Martian sky, two and a half times the full height of Mauna Kea, making it the highest mountain we have ever seen. Astronomers have peered at the greatest mountain in the Solar System since the late nineteenth century, but it wasn't until 1971, when the Mariner 9 space probe took the first images of it, that its true scale was revealed. Just like Mauna Kea, Olympus Mons is a shield volcano. Over millions of years, layer upon layer of lava has built up during long periods of continuous eruptions, slowly raising the mountain to truly gargantuan heights. With a footprint roughly the size of the state of Arizona and gentle slopes leading kilometre after kilometre up to its summit, Olympus Mons is so vast that if you were standing on its foothills it would be impossible to see to the top. It grew so tall because of the specific geology of Mars. In Hawaii, a string of volcanoes is being produced as the Pacific Plate moves over a static hotspot. As the plate drifts northwards over the hotspot, new volcanoes are formed and existing ones become extinct as they are carried away from their source of heat. On Mars, the lack of plate movement over the hotspot beneath Olympus Mons has meant the lava has simply piled up.

Apart from its sheer size, almost everything else about Olympus Mons is familiar. It shares many of the features and qualities we find in shield volcanoes on Earth because the geological processes that built them are identical. Mars and Earth share more than geological similarities, however. The origins of all the rocky inner planets – Mercury, Venus, Earth and Mars – are very similar. The story of their formation can be traced back billions of years to the fiery birth of the Solar System and the planet we call home ◉

BELOW: This colour mosiac taken by the Viking I Orbiter shows Olympus Mons, named after Mount Olympus – the mythical home of the Greek gods. It is the highest volcano on Mars, with a footprint the size of Arizona.

BOTTOM: Olympus Mons is a staggering 550 kilometres (340 miles) in diameter and its 80-kilometre (50-mile) -wide summit caldera is positioned 25 kilometres (16 miles) above the surrounding plains.

MOUNT EVEREST: 9KM

OLYMPUS MONS: 25KM

MANAU KEA: 10KM

SEA LEVEL

ARIZONA

550 KM

DEAD OR ALIVE

LEFT: This artist's animation shows the swirling disc of dust that will build a new planet. It is believed that planets are born small, developing into larger planets as they slowly accumulate dust and gas in bigger clumps. As they grow they collide with proto-planets, until eventually they form a few Earth-sized rocky planets.

THE FORMATION OF ROCKY PLANETS

Around 4.6 billion years ago, the Sun had just ignited and the infant Solar System was nothing more than a disc of gas and dust orbiting around a newly formed star. This proto-planetary disc contained all the matter that would later form the planets and moons of our solar system, a process that would take millions of years of slow construction.

The processes by which solar systems form out of discs of dust and gas surrounding young stars are not fully understood, but the most widely accepted explanation is known as the 'planetesimal' theory. Planets are born small, growing through a gradual accumulation of dust and gas into larger and larger clumps. Dust particles in the disc form small clumps as they collide together randomly over millions of years, and some accumulate more and more mass through these collisions until they reach a critical size of about one kilometre (0.6 miles). These solid, ill-defined objects are known as planetesimals. After reaching this critical size, it is thought that mini-planet-sized objects grow very quickly, because their growth rate increases as their mass increases. This process is known as runaway accretion and lasts only a few tens of thousands of years. There are frequent collisions between the many proto-planets orbiting around the young Sun, but eventually, through a process of continual collisions and mergers, a few Earth-sized rocky planets will be left.

Slowly, these newly formed balls of rock are transformed. A combination of heat from multiple collisions, plus the heat generated by the decay of radioactive elements that were present in the proto-planetary disc, can melt whole areas deep inside the planets. This allows gravity to take over, and so the heavy elements, such as iron and many of the heavier radioactive nuclei, sink to the planet's core ◉

PLANET FORMATION

The Solar System consists of three major types of planet: ice giant, gas giant and terrestrial. These are produced because the protoplanetary disc has different proportions of rock and ice depending on its distance from the Sun. Terrestrial planets develop closer to the Sun where the protoplanetary disc is mainly rock, whilst ice giants develop furthest away from the Sun where the protoplanetary disc is mainly ice.

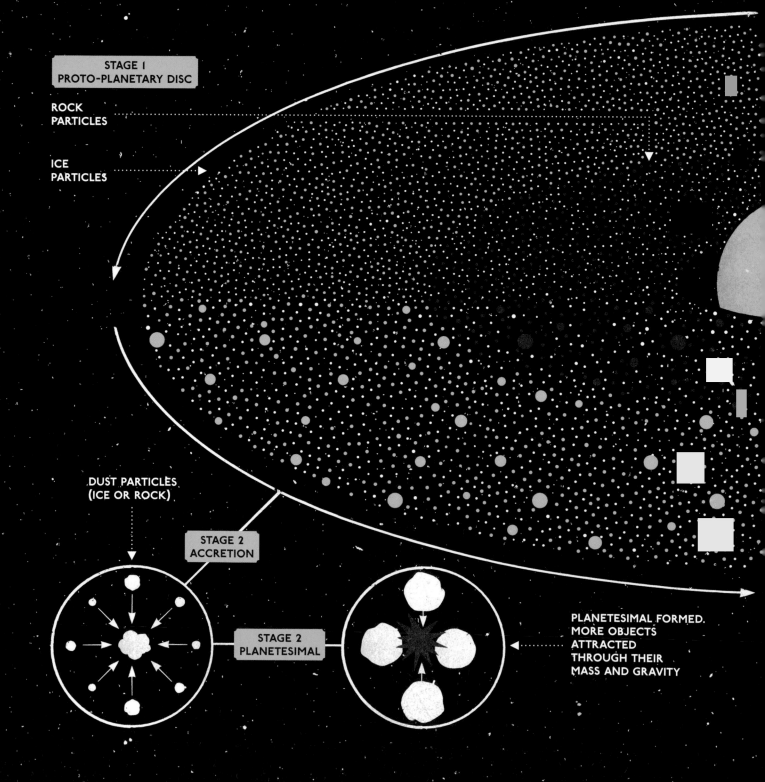

STAGE 1
PROTO-PLANETARY DISC

ROCK PARTICLES

ICE PARTICLES

DUST PARTICLES (ICE OR ROCK)

STAGE 2 ACCRETION

STAGE 2 PLANETESIMAL

PLANETESIMAL FORMED. MORE OBJECTS ATTRACTED THROUGH THEIR MASS AND GRAVITY

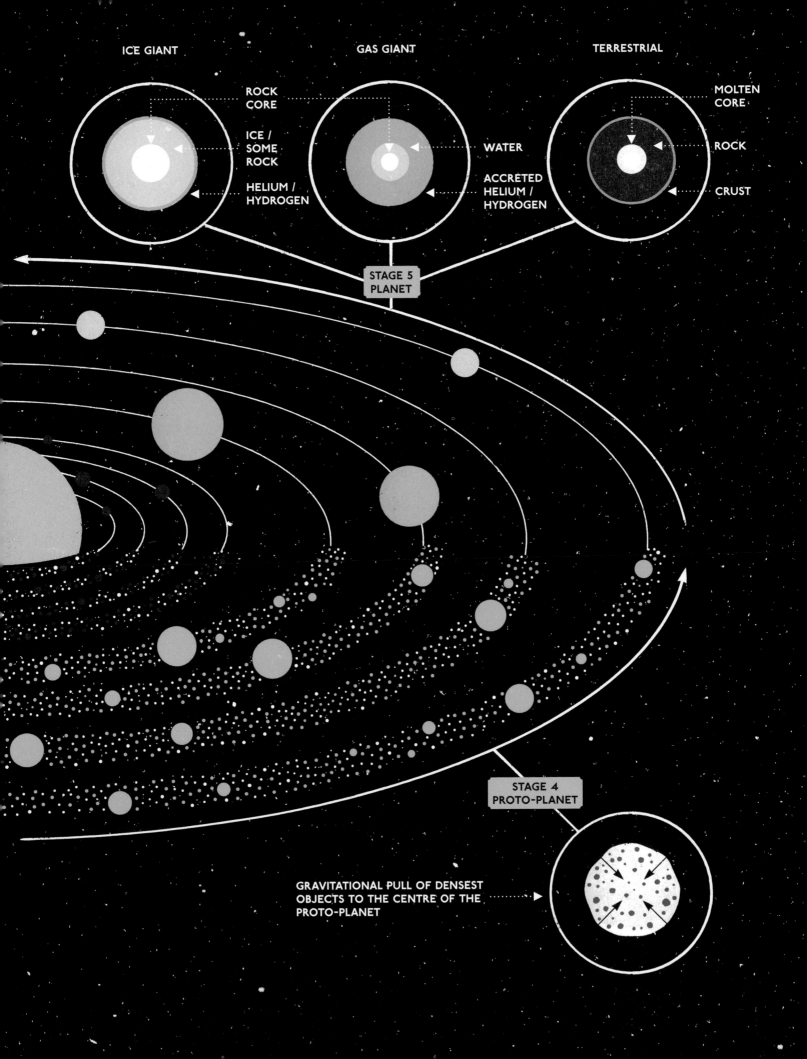

ICE GIANT

ROCK CORE

ICE / SOME ROCK

HELIUM / HYDROGEN

GAS GIANT

WATER

ACCRETED HELIUM / HYDROGEN

TERRESTRIAL

MOLTEN CORE

ROCK

CRUST

STAGE 5 PLANET

STAGE 4 PROTO-PLANET

GRAVITATIONAL PULL OF DENSEST OBJECTS TO THE CENTRE OF THE PROTO-PLANET

Today on Earth, we can still see the remains of this primordial source of heat that has been trapped for billions of years inside our planet's core. It is released in the most spectacular fashion through the eruption of volcanoes. All of the Earth's volcanoes are driven by this ancient source of power, as are the shifts in our tectonic plates that move whole continents and raise great mountain ranges towards the sky. However, elsewhere in the Solar System this powerful source of energy ran out long ago.

The volcanoes on Mars are little more than a petrified memory of a distant, more active past. For all its grandeur, Olympus Mons stands cold and extinct, and when we look down across the rest of the surface of Mars, we see no evidence of any kind of geological activity. As far as we can tell, Mars is now a dead world; its geological heartbeat has been extinguished. Despite having the biggest volcanoes in the Solar System, the primordial heat that formed them is no longer beneath the Martian surface. Something has stopped the red planet in its tracks.

LEFT: Eruption of Pu'u O'o Crater, Hawaii, 1985.

NEWTON'S LAW OF COOLING

Space is cold, very cold. The temperature of the Universe is on average just over 2,72 Kelvin – around -270 degrees Celsius. This is very close to absolute zero. The fact that the Universe isn't at 0 Kelvin is significant, because this precisely known number is the background radiation left over from the beginning of the Universe; a fading echo of the Big Bang 13.7 billion years ago.

In this freezing heat bath hotter objects, including planets, lose heat to space. This is not lost by convection or conduction, since space is almost a vacuum, instead planets lose their heat through radiation – the emission of infrared light. The overwhelming majority of this energy radiated into space is simply the energy a planet receives from the Sun. If Earth didn't re-radiate the Sun's energy away at the same rate at which it received it, it would rapidly heat up.

Earth's internal heat source plays an important role; the primordial heat left over from its formation and the radioactive decay of elements deep within its core. The rate of loss of this heat is determined by the ratio of a planet's surface area to its volume, because the internal heat must be radiated out into space from its surface.

This is the key to understanding why Mars is now geologically dead. Mars is about half the diameter of Earth and just one-eighth of its volume. A bit of maths explains this: volume is proportional to the cube of the diameter (volume is measured in cubic metres, the diameter of a sphere in metres), so Mars would have stored less internal heat initially because it is smaller. The critical factor is the surface area available to radiate the heat away. Surface area is proportional to the square of the diameter (measured in square metres), so Mars has one-quarter of the surface area of Earth but only one-eighth of the volume. This means it has more surface area in relation to its original heat store and so lost its inner heat much faster.

The combination of these two factors defines the life of a planet. Millions of years ago when the interior of Mars grew cold, the volcanoes lost their life blood, the geological heart of the planet died and its surface ground to a halt. The fate of a whole planet was destined by the simplest of laws of physics and the unstoppable flow of heat ◉

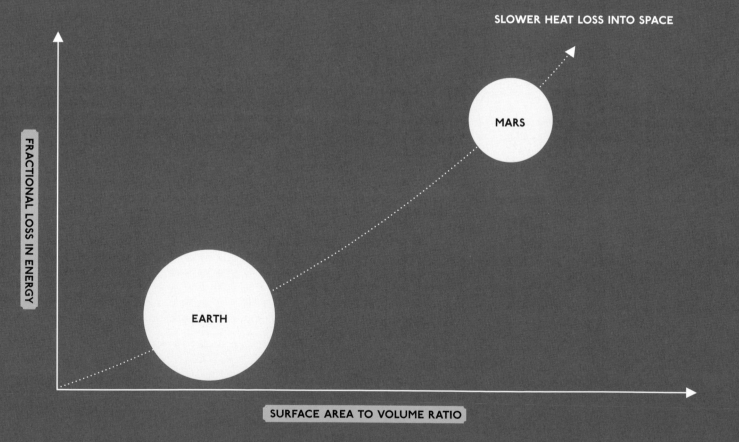

PLANETARY HEAT LOSS IN SPACE
As the Earth is eight times the volume of Mars, it has more primordial heat. It also has four times the surface area, thus loses heat more slowly. The combination of these two factors results in a warm Earth and a geologically dead Mars.

VENUS: A TORTURED HISTORY

Here on Earth, we have seen one beautiful manifestation of how simple laws of nature play out and build a planet. On Mars, we have another example of what happens when you take a planet smaller than Earth: it loses its heat more quickly and becomes geologically inactive. In the cosmic laboratory of our solar system these are just two examples of the delicately balanced processes that decide a planet's fate, but we also have another nearby planetary experiment to explore. There is a planet just like Earth, but that is positioned a little closer to the Sun; it is the brightest point of light in our night sky, so similar in size to our own world that it has been called Earth's twin.

However, Venus is a tortured world. With an average surface temperature of 464 degrees Celsius, it has the hottest surface of any place in the Solar System other than the Sun. Standing on Venus, you would not only be roasted but also crushed by an atmospheric pressure of over ninety times that of the Earth. The clouds of sulphuric acid above your head would threaten you with rain that would never fall on you because the heat of the planet would evaporate it before it reached the ground. No human could ever stand on the surface of Venus today, but wind back a couple of billion years and the hellish planet may not have been so inhospitable. In its early history, Venus was probably not such a foreboding place.

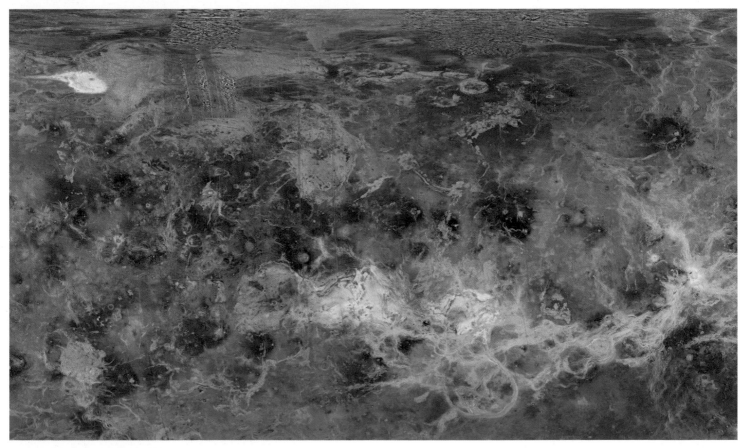

TOP: This is a composite image of the complete radar image collection obtained by the Magellan mission. Launched aboard the space shuttle Atlantis in May 1989, the Magellan spacecraft began mapping the surface of Venus in September 1990.

ABOVE: Venus is an inhospitable world whose surface temperature prevents the existence of life. It is a planet of volcanoes, with over 1,600 located on its surface, including Sif Mons, seen in the background here.

The Deccan Traps in India is a lush green expanse of hills, many with a particular stepped shape (hence the name 'deccan', Dutch for stairs). Today this rich landscape extending for over 500,000 square kilometres (193,000 square miles), but hidden beneath the green is a secret that holds a tantalising clue to understanding how Venus choked to death.

The Deccan Traps are one of the largest volcanic features on Earth. Sixty-five million years ago this area of central-west India witnessed a series of colossal eruptions that lasted for at least thirty thousand years. At one point an area the size of half of modern India was covered by lava, a staggering 1.5 million square kilometres (600,000 square miles). The impact on the Earth's climate was equally enormous; millions of tonnes of volcanic ash and gases were hurled into the atmosphere with a devastating effect on life. These eruptions affected the climate so profoundly that it is possible they played a role in the mass extinction events at the end of the cretaceous period that wiped out over two-thirds of the species on Earth.

It's almost impossible to imagine how these colossal eruptions must have looked when you visit the Deccan Traps today, but despite the tranquil appearance of the verdant-stepped hills, this place was created from one of the most sustained and violent events our planet has ever known, and the formation of this landscape is echoed on the surface of our nearest planetary neighbour.

Until the Magellan probe arrived in 1990, the thick layer of opaque cloud that surrounds Venus severely limited our knowledge of the planet's surface. With the radar-mapping equipment aboard this pioneering probe, we could peer through the cloud and create the first, and to date, best images of the hidden landscape below. Magellan saw a world of volcanic destruction way beyond anything seen on Earth. It is a landscape built on exactly the same geological foundations as the Deccan Traps, but on a far larger scale.

We have discovered over 1,600 volcanoes on Venus' surface, far more than on any other planet in the Solar System. At least 85 per cent of the planet is covered in basalt lava plains that have poured across the surface. It is not know for certain if Venus is still volcanically active, but because it is a similar size to Earth it might be expected still to have a hot geological heart powering its volcanoes. As yet we haven't directly witnessed any eruptions, but there are clues that suggest volcanoes have been active in the relatively recent past. The Magellen probe spotted ash flows near the summit and north flank of Venus' tallest volcano, the eight-kilometre (five-mile) -high Maat Mons, and more recently, in 2010, the ESA spacecraft Venus Express provided evidence of volcanic activity occurring as recently (in geological terms) as 2.5 million years ago, and maybe even much later.

Many of the volcanoes on Venus are identical to the ones found on Earth. Shield volcanoes like Maat Mons litter the surface, but although the underlying geology is the same, the character of these volcanoes can be very different. On Earth, shield volcanoes like Mauna Kea and the Hawaiian volcanoes can be up to ten kilometres (six miles) high but much wider, but on Venus some volcanoes have footprints of hundreds of square kilometres but average heights of only 1.5 kilometres (1 mile). The Venetian volcano Sif Mons is a massive 300 kilometres (186 miles) across but only 2 kilometres (1.2

BELOW: These three circles signify volcanoes located in the Guinevere Planitia lowland on Venus. The central volcano appears to be very round with steep slides and a flat top.

BOTTOM: This aerial view clearly shows approximately 200 small volcanoes scattered over the surface of Venus. The volcanoes range in diameter from two to twelve kilometres (one to seven miles).

miles) high. Venus also has a type of volcano that doesn't exist on Earth. The 'tick' volcanoes were so-called because of their resemblance to the insect and are thought to be the remains of volcanic domes that have collapsed. Another of Venus' oddities are the thousands of strange flat volcanoes called pancake domes that are clustered in groups across the Venetian surface and are much wider than any similar structures on Earth. Venus truly is a world dominated by volcanoes, but unlike on Earth, this intense geological activity pushed our cosmic twin down a path of no return.

Four billions years ago it is thought that Venus was a world much like our own. The climate was cooler and the surface may have been covered in vast oceans of water. Just like our own planet, this wet warm environment may have been the perfect place to harbour life. How long these conditions lasted for is still unknown, but there is evidence suggesting that Venus was a more welcoming place for up to two billion years. If this was the case, then many scientists believe Venus could have been the most likely place for life to evolve beyond our own planet. If the conditions were stable enough for a few hundred million years, then life may even have flourished before the character of the planet turned ugly.

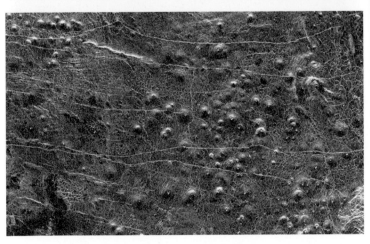

This is one reason why understanding the history of Venus is so important. This hellish world provides the most telling example our solar system has to offer of the potential fragility of planetary environments.

Explaining why Venus and Earth reacted so differently to the same kind of volcanic cataclysm requires an understanding of a matrix of different factors. Volcanoes don't just produce heat and lava, they also produce vast amounts of greenhouse gases like carbon dioxide. Every planet, including Earth, absorbs energy from the Sun as visible light. This light streams through our atmosphere almost untouched and is absorbed by the ground, heating it up day after day. The ground then re-radiates this energy as infrared radiation. Atmospheric gases, particularly carbon dioxide, are very good at absorbing infrared light, and so they trap the heat and the planet heats up. The more greenhouse gases in the atmosphere, the more a planet will heat up.

On Earth, we are beginning to see the effect that an increase in greenhouse gases created from the burning of fossil fuels has on our climate. Global warming is a phrase that has only recently entered popular vocabulary, but it would have wreaked havoc on our planet long ago if it hadn't have been for a familiar characteristic of our weather.

One of the most important reasons we have taken such a different path to Venus is something that happens so often on Earth that we take it for granted. Rain plays a significant role in keeping our planet a pleasant place to live. It acts as part of a global recycling system, keeping our atmosphere in balance by washing out potent greenhouse gases like carbon dioxide and locking them away in rocks and oceans. On Venus, the planet's position in the Solar System, combined with the laws of physics, have conspired to make it impossible for rainfall to cleanse its atmosphere. Because it is slightly closer to the Sun, and so a little hotter than Earth, Venus lost all its liquid water. The oceans of Venus would have gradually evaporated into the atmosphere. The potentially life-giving water would have simply escaped off into space.

With no water, there is no rain on Venus, and so for billions of years there has been nothing to temper the build-up of volcanic gases in its atmosphere. Venus ended up cocooned in a thick, dense, high-pressure blanket of greenhouse gases, making the temperature inexorably rise and turning Venus into the hell-like world we see today.

Compared to scorched Venus and frozen Mars, our planet is a very special ball of rock. Although governed by the same universal set of rules, the Earth is not too big, not too small, not too hot and not too cold. This is why Earth has been called the 'Goldilocks planet', because everything seems just right, but the life and death of our planet is influenced by more than just the forces emanating from the depths of our own world. Our fate is intimately connected with our cosmic neighbours in ways that are subtle and complicated but extremely powerful ◉

JUPITER: KING OF THE GIANTS

Jupiter, King of the Gods, the fifth planet from the Sun, has been revered since ancient times. Visible to the naked eye in the night sky, it can also be seen during the day when the Sun is low on the horizon. For millennia, humans have looked to Jupiter and imbued it with power. From the Romans to the Greeks, the Chinese to the Hindus, almost every civilisation on Earth has gazed at its light without realising the true influence that it exerts over our solar system.

RIGHT: An artist's concept of *Pioneer* over Jupiter's Red Spot

JUPITER: THE ETHEREAL PLANET

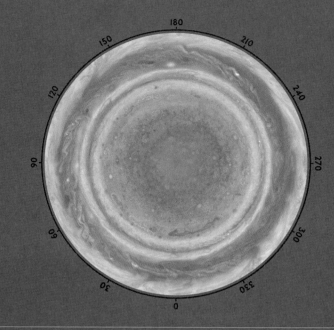

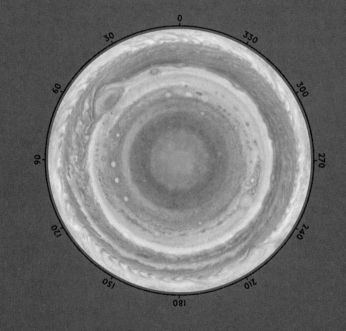

Jupiter is by far the largest planet, so big you could fit the Earth inside it over 1,000 times, and it is of a completely different character to the inner rocky worlds. It's one of the four gas giants that circle the Sun, and along with Saturn, Uranus and Neptune, it is made up of the same stuff as a star – hydrogen and helium, the most common elements in the Universe. Although it may have a solid core made up of heavier elements, Jupiter, like all the gas giants, is an ethereal planet, a planet with no real boundaries between its skies and no substance further down. It is a vast atmosphere that gets denser and denser as you travel deeper. Despite its seemingly insubstantial nature, though, Jupiter is truly a massive planet. It has a mass that is two and a half times that of all the other planets put together. It is so big, theoretical models suggest that if it were any more massive, it would begin to collapse further under its own gravity, transforming into a sub-star-like object called a Brown Dwarf. Jupiter is probably about as big as a planet of its composition and construction can be, and that means it dominates the rest of our solar system.

Astrologers have long claimed that Jupiter can influence our lives, but we now have scientific evidence that this mighty planet does indeed have a significant connection with our own small world, although not in the way that the ancients thought. Despite the fact that, as Sir Patrick Moore famously said, astrology proves only that there is one born every minute, Jupiter influences our planet across more than half a billion kilometres of space through the force of nature that binds the galaxy together.

Gravity is one of the four fundamental forces of Nature. It shapes so much of our universe and yet it is by far the weakest force, a force that we can resist with ease. As stated previously, pick a rock up off the ground and you are defying the force of an entire planet. Despite its weakness, gravity does have two properties that allow

ABOVE: These maps were constructed using images produced by NASA's Cassini spacecraft. They are the most detailed global colour maps of Jupiter ever produced and use colours that would be close to those that the human eye would see when viewing Jupiter. The maps show views of both poles of the planet, featuring colourful clouds, parallel reddish-brown and white bands, the Great Red Spot and blue-grey areas that denote 'hotspots'.

OPPOSITE: This sequence of nine images shows Jupiter as it rotates through more than a complete 360-degree turn. This massive planet rotates more than twice as fast as Earth, completing a single rotation in around ten hours. Its powerful gravitational force directly influences every other object in the Solar System.

it to shape our universe. Everything that has mass (or energy) attracts everything else, and if you add more mass to something the gravitational force between it and other objects increases. It also has an infinite range, which means that its influence can stretch across the entire Solar System and beyond. Gravity never quite goes away, so you can be a long way from its source and still feel its effects.

As the force of gravity is directly related to the mass of an object, because it is the most massive planet Jupiter has the most powerful gravitational field in the Solar System other than that of the Sun. It is this gravitational force that directly influences every other object in the Solar System. Jupiter's gravitational pull is strong enough to profoundly influence the orbits of passing interplanetary asteroids and other wandering space debris, even at large distances away.

This effect on the wandering stuff of our solar system can play out in three different ways. Firstly, Jupiter can capture it, literally pulling it inwards on a collision course and ultimately merging it with the gas giant itself. Secondly, it can change its orbit around the Sun in such a way that it throws it out of the Solar System forever. The third of Jupiter's options as it marshals the solar traffic is perhaps the most worrying one for us today. If the angles are just right, the planet can deflect an orbiting asteroid into a new orbit, and occasionally place it on a direct collision course with the rocky inner planets, including our own ◉

ARMAGEDDON WATCH

BELOW: At the top of the mountain of Heleakala, Hawaii, is the technological solution to the problem of detecting dangerous asteroids on course to Earth. The telescope contains one of the largest digital cameras ever built, designed to capture images of 1,400 megapixels.

RIGHT: This meteorite is a sample of the crust of the asteroid Vesta. This is only the third object collected from the Solar System beyond Earth (the other two are Mars and the Moon). This unique meteorite is almost entirely made of the mineral pyroxene, which is common in lava flows.

This telescope, known as the PS-1 Observatory, sits on top of the mountain of Heleakala on the Hawaiian honeymoon island of Maui, and it could one day save your life. It contains some of the largest digital cameras ever built, designed to capture images with a staggering 1,400 megapixels (1.4 billion pixels) on an area about forty square centimetres (six square inches). To put that into context, a good domestic digital camera contains about ten megapixels on a chip just a few millimetres across. The telescope is designed with one purpose in mind – to hunt down killer asteroids. The threat is easy to state: if anything bigger than a kilometre in size hits the Earth, it would probably kill almost everyone on the planet. This is the reason that on most nights, as you sleep soundly in your bed, this revolutionary camera is scanning vast swathes of the sky looking for signs of an approaching apocalypse.

Most of the known asteroids within our solar system orbit in the asteroid belt that sits between Mars and Jupiter. We don't know for certain how many objects sit in this belt,

> *If anything bigger than a kilometre in size hits the Earth, it would probably kill almost everyone on the planet.*

but over 200 are known to be larger than 100 kilometres (60 miles) across and a few million bigger than a kilometre. That sounds like a lot, but in fact the asteroid belt is mostly empty and no spacecraft we have sent through it has ever encountered a problem.

One of the objects in this region, named Ceres, is so big that it was classified as the eighth known planet when it was

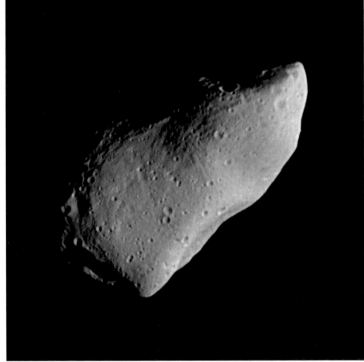

BELOW: This spectacular image of asteroid 951 Gaspra was captured by the Galileo spacecraft in 1991 when it made the first close flyby of an asteroid. The blue patches are believed to be fresher rock than the older, reddish areas . The asteroid is about 19 by 12 by 11 kilometres (12 by 7 by 7 miles) in size.

PLANET OR ASTEROID? Until 2006, Ceres was considered the largest asteroid in the main asteroid belt. However, it has now been reclassified as a dwarf planet because, unlike the other asteroids, it can become spherical under its own gravity.

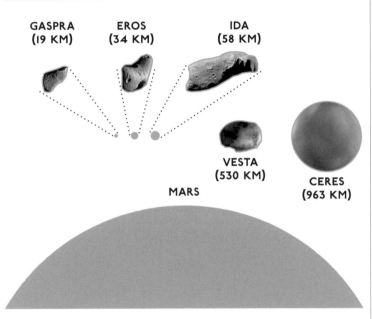

GASPRA
(19 KM)

EROS
(34 KM)

IDA
(58 KM)

VESTA
(530 KM)

CERES
(963 KM)

MARS

discovered in 1801. When other rocky bodies were found in the same area scientists realised Ceres was just one of many, and so by the middle of the century it was relegated to the new status of asteroid (meaning star-like) coined by William Herschel. It wasn't until 1991 that we got close enough to the asteroid belt with the Galileo spacecraft to take an intimate look at one of these mini-worlds.

The orbit of most of these millions of asteroids takes them around the Sun without any threat to Earth. But as well as these predictable asteroids, there is another type that poses a far greater risk. We currently know of over seven thousand Near-Earth Asteroids, with nearly a thousand of these being bigger than one kilometre (0.6 miles). These are asteroids with orbits distorted from the harmless majority that stay away from the inner solar system, bringing them near enough to Earth for concern. We have catalogued thousands of these, but nobody can be sure how many are out there, or what effect a subtle change in the orbit of one we know about would be. Thus, the work of the PS-1 Observatory and other lookouts

across the planet are crucial for our future safety.

Each night the observatory team are looking for any unidentified objects that might be heading our way. Any point of light could be an asteroid in an orbit that brings it perilously close to Earth, but spotting the rocks from the stars is not an easy process. To help them to make as accurate an analysis as possible, the camera at the PS-1 Observatory captures several images of the same patch of sky, taken minutes apart. The team can then see if anything has moved, relative to the backdrop of stars. By literally subtracting the images from each other, anything that stands still – i.e. stars – will disappear, but anything that has moved in the time between the two photographs will still be there. Fast-moving bright objects are all that will remain in the photographs.

In the whole night sky we may well be able to detect hundreds of objects that we have never known existed. Many of these menacing lumps of rock are in eccentric orbits that bring them close to Earth – and all because at some point in their lives they came under the influence of Jupiter's gravity ◉

If you ever needed a demonstration of how congested space is near the Earth, just look at the picture opposite. Every one of those points of light is an asteroid that we know of, and Earth is swimming right through the centre of them. So when you look up into a nice clear night sky for reassurance that we are safe, remember this picture, remember that our planet always has been and always will be trapped in a deadly game of dodge ball – in a game where the gravitational stranglehold of Jupiter regularly throws asteroids our way.

RIGHT: The sky is now being methodically scanned for asteroids that might cross Earth's orbit (the blue streak in the time-lapse photograph right). The complex interaction between the gravitational pull of the Sun and the other planets (particularly Jupiter) and these near Earth objects means it is very difficult to calculate precisely their final trajectories and, as some of them pass between the Earth and the Moon, the margin for error is quite small.

COLLISION

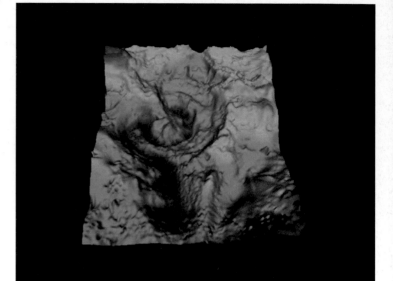

BELOW: One of the world's most famous impact sites is the Barringer Crater in Arizona. Around 50,000 years ago, a 300,000-tonne, 50-metre (30-feet) in diameter lump of iron and nickel entered the Earth's atmosphere, making this crater.

RIGHT: This computer-generated image is a gravity map of the Chicxulub Crater discovered on Mexico's Yucatan peninsula, which shows the crater is made of many rings, including an outer one of 300 kilometres (186 miles) in diameter.

When geophysicist Glen Penfied began searching for oil in the Yucatan peninsula of Mexico in the late 1970s, he had no idea of the discovery that was lurking beneath his feet. Penfield was surveying this area in order to pinpoint new locations to begin drilling in, but his interest was quickly diverted to a different kind of geological treasure. The geophysical data Penfield began to unearth suggested that hidden within this area was an impact crater of vast proportions. The evidence suggested that a catastrophic impact had taken place here that created a crater that is over 180 kilometres (112 miles) wide.

Today, this extraordinary feature is known as the Chicxulub Crater. Named after the town at its centre, the crater has been studied for over twenty years by hordes of experts. It is one of the largest-known impact craters on the planet and it is estimated that the object that struck here was at least ten kilometres (six miles) across. The scale of this impact alone makes it an extraordinary location, but the timing of the impact is what has elevated Chicxulub into the A-list of asteroid sites.

The asteroid that struck here is thought to have slammed into the Earth 65 million years ago at the end of the Cretaceous period. It coincides perfectly with the most famous extinction event in the history of the planet – the mass extinction event that caused the disappearance of the dinosaurs. Although there is no complete agreement amongst the scientific community about this link, the overwhelming consensus is that the Chicxulub impact was the trigger for the extinction of the largest creatures ever to walk the Earth.

We may never know for certain where this enormous asteroid came from or what set it on its course to Earth, but we are pretty sure it originated in the heart of the asteroid belt between Mars and Jupiter. Some scientists have suggested that the dinosaurs' fate was sealed by a collision in the asteroid belt that created a family of asteroids, with one in particular that headed for Earth. What is certain is that wherever the asteroid came from, its journey to Earth was influenced by the mighty presence of Jupiter. Jupiter's gravitational influence on passing space debris has made our planet a world under constant bombardment. Earth is littered with impact sites, from the most visually stunning and famous, such as the Barringer Crater in Arizona, to the hidden craters that have dissolved from our view over billions of years ◉

JUPITER'S GRAVITATIONAL KICK

The asteroid belt is a vast expanse of space extending over 240 million kilometres (150 million miles) between Mars and Jupiter, further than the distance from the Earth to the Sun.

Now and again, because of collisions in the asteroid belt, a stray asteroid will get thrown into a position where it periodically aligns with Jupiter over and over again and settles into a rhythm known as orbital resonance. Jupiter is such a massive planet that it will give such an asteroid a gravitational kick, changing its orbit. Over time these orbits can become elongated or elliptical rather than circular, which means that they can get thrown into the inner solar system and cross the orbits of the inner planets, including that of the Earth.

Jupiter was once thought to be our protector, its enormous gravity swallowing up dangerous asteroids, but we now realise its gravitational influence can actually propel some of those asteroids in our direction, creating the huge craters we see in places like Chicxulub. Although the idea of these impacts appears to be purely destructive, the surprising thing is that these catastrophic events may actually have been instrumental in shaping our planet and the life that has flourished on it since. Impacts have been one of the driving forces of evolution on Earth, changing the climate and triggering extinctions – when huge swathes of life on Earth are wiped out, it creates the ecological niches into which other species like ourselves evolve.

It is incredible to think that a planet located more than half a billion kilometres away could dictate our fate and define the life and death of a whole world ◉

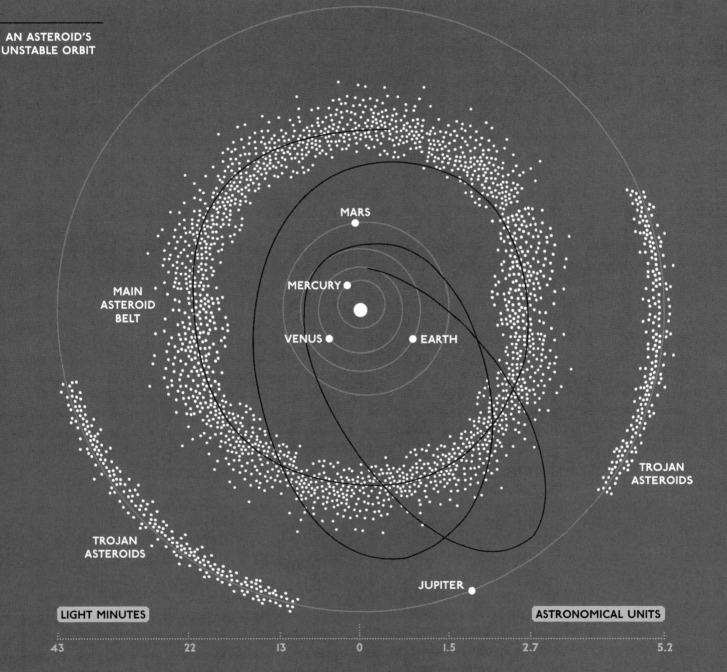

AN ASTEROID'S UNSTABLE ORBIT

MARS

MAIN ASTEROID BELT

MERCURY

VENUS · EARTH

TROJAN ASTEROIDS

TROJAN ASTEROIDS

JUPITER

LIGHT MINUTES

ASTRONOMICAL UNITS

43   22   13   0   1.5   2.7   5.2

POWERFUL CONNECTIONS: THE MOON AND TIDES

BELOW AND OPPOSITE: The Minas Basin, in the Bay of Fundy, Nova Scotia, Canada, experiences the most dramatic differences between low and high tides on Earth. Typically the high tide (shown opposite) at the head of the Bay of Fundy can be as much as seventeen metres (fifty-six feet) higher than at low tide (shown below). These large tides are the result of tidal resonance, which can become synchronised with the lunar tides, amplifying its effect.

Our understanding of the Solar System began much closer to home. Gazing down at us, it was our moon, with its regularly changing face, that first fired our interest in worlds beyond our own. When we could look further out, we discovered the Solar System was full of moons, each invisibly connected to their parent planets by gravity.

Every year on 18 August, thousands of people trek to the Quiantang river in south-eastern China to witness one of the great spectacles of the natural world. The river and bay are famed as the location of the world's largest tidal bore, a phenomenon that creates a wall of water up to fifty metres (thirty feet) feet high, travelling at forty kilometres (twenty-five miles) per hour. This true tidalwave crashing along the Quiantang river is a spectacular reminder of one of the most powerful effects on Earth. Tidal bores occur in only a handful of places as they need both a particular shape of river system and a large tidal range, but these rare events are just one extreme example of something that many of us witness every day without giving it a second thought.

Across our planet tides rise and fall every twelve hours

THE EFFECT OF THE MOON ON EARTH'S TIDES

High tides occur at the point where the gravitational pull on Earth's oceans is strongest, pulling the water towards it and raising the level of water. Spring tides occur when the Sun and Moon are aligned.

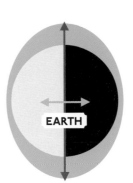

3RD QUARTER MOON

NEW MOON

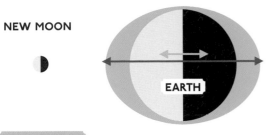

EARTH

SPRING TIDES

NEAP TIDES

EARTH

and twenty-five minutes. They are the most visible mark of the most intimate interaction we have with any celestial body. Stand on a beach and watch the tide ebb and flow and you are witnessing the direct effect of the Moon on the body of water that covers our planet. On one side of the planet the high tide occurs at the point on Earth where the ocean is closest to the Moon, and so the gravitational force is at its strongest, pulling the water towards it. Twelve hours later and this same location will be at the furthest point from the Moon and so the gravitational pull on the water in the ocean is at its weakest. At this point the Earth is pulled towards

the Moon slightly more than the water, and there is a high tide at that point. The Sun can also affect the tides on Earth through the same differential gravitational effect, but even though the Sun is significantly more massive than the Moon, it is further away and so the effect is much weaker. It is our lunar companion that most powerfully drives this bi-daily rise and fall of the oceans. We've been studying the tides and plotting their rhythms for thousands of years, unaware that elsewhere in the Solar System an identical relationship between a moon and a planet has created a much more extreme tidal phenomenon ◉

JUPITER'S MOONS

BELOW: Ganymede is one of Jupiter's four largest moons, completing an orbit of the planet every seven days. It is also the largest moon in the Solar System – even larger than the planets Mercury and Pluto and just over three-quarters of the size of Mars.

OPPOSITE: In this image of Europa taken by NASA's Galileo spacecraft, the colours have been enhanced to show the differences in materials that cover the moon's icy surface. The red lines show cracks and ridges, thousands of kilometres long, which are caused by the tides raised by the gravitational pull of Jupiter.

Gravity is a two-way street. Isaac Newton put it in more scientific terms; to every action there is an equal and opposite reaction. So not only does the Moon exert a force on the Earth, but the Earth exerts an equal and opposite force on the Moon. Over hundreds of millions of years, this gravitational embrace between Earth and Moon has had a profound effect. We only see one side of the Moon facing Earth because the Moon is 'tidally locked'; its rotation rate matches its orbital period around the Earth. This is no coincidence but a consequence of the gravitational interaction between the two bodies. However, on a daily basis the impact of Earth's gravity is minimal. There is no liquid water on our moon and so there are no ocean tides, and Earth's gravity is too weak to have a significant effect on the Moon's rocky constitution.

Half a billion kilometres away, we've discovered a very different story. The powerful gravitational bond that exists between one moon and its parent planet, Jupiter, has done something astonishing; it has brought the moon to life, making it the most violent place in the Solar System.

Four hundred years ago, Galileo was the first human to turn his telescope to the night sky and look at Jupiter. He immediately noticed that the giant planet was not alone. On 7 January 1610, Galileo observed three points of light around Jupiter. He described them at first as 'three little stars', but over the next few nights he quickly realised that they moved in relation to the planet, disappearing from view and reappearing. He correctly surmised they could not be stars; they must be objects orbiting in the Jovian system. By 13 January, Galileo had observed and catalogued the four largest moons of Jupiter, and in doing so confirmed Copernicus' revolutionary view of the Solar System. No longer did our view of the Universe adhere to the Aristotlean one that all heavenly bodies must orbit the Earth. Here was direct evidence of other worlds orbiting another planet, unequivocally breaking the divine symmetry of the Earth-centred cosmos for good and challenging what for many was a deeply held belief.

Jupiter's four largest moons are named after the lovers of the Greek god Zeus. Furthest out is Callisto, a ball of rock and ice the size of Mercury and the third-largest moon in the Solar System. Next is Ganymede, the largest moon in the Solar System, which is the only moon known to have its own internally generated magnetic field, and which may harbour a saltwater ocean deep below its surface. Next is icy Europa, the smoothest, most tantalising moon. Its surface is crisscrossed by dark streaks and gathered evidence suggests there is a vast ocean below its surface. For many scientists, Europa is now the most likely candidate to harbour extraterrestrial life. Finally, closest to Jupiter is the small yellow-tinged moon, Io. Modern space probes have revealed that Io is an incredibly tormented world; somewhere we can glimpse by visiting one of the most inhospitable places on Earth ◉

	IO	EUROPA	GANYMEDE	CALLISTO
Discovered	1610	1610	1610	1610
Mass (Earth = 1)	1.4960e-02	8.0321e-03	2.4766e-02	1.8072e-02
Equatorial radius (Earth = 1)	2.8457e-01	2.4600e-01	4.1251e-01	3.7629e-01
Distance from Jupiter (km)	421,600	670,900	1,070,000	1,883,000
Orbital period (days)	1.77	3.55	7.15	16.69
Orbital velocity (km/sec)	17.34	13.74	10.88	8.21

JUPITER

(Not to scale)

ERTA ALE,
NORTH-EASTERN
ETHIOPIA

In the Afar region of north-eastern Ethiopia stands one of the rarest geological phenomena on our planet. Erta Ale is the most active volcano in Ethiopia, and at just 610 metres (2,000 feet) high it is one of the lowest volcanoes in the world. But what makes this volcano special are the lava lakes that have continuously dominated its summit for over a century. Lava lakes are incredibly rare; there are currently only five sites on Earth where they can be seen, but none have existed as long as those on the 'smoking mountain' of Ethiopia.

Erta Ale was by far the most challenging place we visited during the filming of *Wonders of the Solar System*. It sits in the Danakil Depression, the remote and hostile region of north-east Africa where the Great Rift Valley meets the Red Sea. The region is intensely geologically active because it is situated at the Afar Triple Junction, a delicate place in the Earth's crust where the Red Sea and Gulf of Aden meet the East African Rift. The Earth's crust is literally being ripped apart, leading to intense earthquakes and volcanic activity. Even as I write these words from the vantage point of a year, I find recalling this adventure both evocative and exciting. The Great Rift Valley is our birthplace – we are all related to someone who lived in the place we now call Ethiopia. In a remarkable piece of research based on the human genome project, it has been shown that human genetic diversity declines steadily with distance from

Addis Ababa, Ethiopia's capital city. In other words, we began the long march across the globe from the region around Addis. Ethiopia itself as a geopolitical entity is Africa's oldest independent nation, with a rich history stretching back well over 2,000 years, but a great civilisation existed in this region many hundreds if not thousands of years before that. You cannot visit Ethiopia without glimpsing in your peripheral vision a line of ghosts standing by your shoulder, winding back through the ages to the birthplace of our species.

We began the intrepid part of our journey from a military airfield in the northern city of Mek'ele. The machine charged with ferrying our film crew to Erta Ale was an ageing but reassuringly rugged-looking Russian Mi8 transport helicopter; a reliable workhorse, I was told – there are more Mi8s flying than any other type of helicopter in the world.

The approach to Erta Ale from the air was unusually bleak and quite daunting. The landscape is lunar – although more desolation than magnificent desolation. It is an unremitting expanse of slate-grey basalt and baked brown rock, drained of colour by the brutal Sun. To protect us from the 'smoking mountain' were a dozen Afar tribesmen, the nomadic people whose permission and protection are essential for a visit to Erta Ale. To the locals, the volcanoes largest lava lake is known as the Gateway to Hell.

ALL IMAGES: We were flown out by military helicopter to the farthest, most inhospitable reaches of north-east Africa. There we saw one of the rarest geological phenomena on our planet – a volcano with a lake of molten lava. The volcano has been named Erta Ale by local people, which means 'smoking mountain'.

DEAD OR ALIVE

LEFT: Erta Ale, in the Afar region of
Ethiopia, is surrounded completely
by an area below sea level. Volcanoes
with lava lakes are very rare.

Erta Ale's lake of lava is a mesmerising sight, especially
at night. The vertical edge of its crater is illuminated by the
bright red glow of liquid rock. The surface of the lake itself
is mostly dark, because the lava quickly cools as it meets the
air, but it is crisscrossed by a series of almost perfectly drawn,
jagged red lines. The reason for these strangely shaped lines is
unknown. Staring at the lake is addictive, because every now
and then a violent mini-eruption occurs somewhere on the
surface of the lava, throwing molten rock vertically upwards
into the night sky and clearing a hole in the darker crust to
reveal the bright red lava beneath. This burst of activity is
accompanied by a bubbling, sloshing sound, a rapid increase
in brightness and occasionally a caustic wave of gases that
instantly burn the throat. This is the signal to grab a gas mask
rather than turn and run, because it is virtually impossible
to avert your gaze from the mountain when its anger rises.
I think anger is the correct word; we developed a respect,
already possessed by our Afar companions, for Erte Ale. It has
a presence that is very difficult to put into words; alive would
be too strong, but its unpredictable power lies somewhere
in the shadowed spaces between animate and inanimate.
I understand absolutely why the Afar believe that demons
emerge from the depths to drag unwary travellers to the
Ethiopian equivalent of Hyades. And we had to camp beside
it for three nights.

Erta Ale is a window into our planet's history – a portal
not only into its depths, but backwards in time – a slopping
and gasping reminder of our Earth's formation.

The magma rises up from many kilometres below the
Earth's crust, circulating to the surface and sinking back
down again. It is a native and vital anachronism left over from
the birth of a large rocky planet close to the Sun, yet we have
seen something similar in the far reaches of the Solar System ◉

THE MOST VIOLENT PLACE IN THE SOLAR SYSTEM

BOTTOM: Jupiter's moon, Io, is the most geologically active body in the Solar System today and provides the most extreme example of the effect of tidal forces. Here we see Pele erupting.

In 1979, the Voyager spacecraft became the first mission to study and photograph the moons of Jupiter close up, including innermost Io. Io is approximately the same size as our own moon, and for 400 years we'd assumed it was a cold dead world, although the Voyager mission scientists knew that expectations are often dashed in a spectacular way when new worlds are visited.

How exciting, then, must it have been to stand in mission control when the first close-up pictures of Io arrived back at Voyager's home planet.

Not everyone was quite so in the dark about what Io might have in store, however. Just weeks before Voyager arrived at Jupiter, three scientists made a prediction that appeared to many planetary scientists as nothing more than fantasy. Using the same physics that underpins our understanding of tides on Earth, they predicted that Io should have an intense internal heat source because of its unique position in the Solar System. Io orbits very close to massive Jupiter – about the same distance from its parent planet as our own moon orbits Earth. It is also surrounded by its large sister moons, Europa and Ganymede, which are orbiting further out. This configuration of moons and planets means Io is under the influence not just of the massive gravitational pull of Jupiter but also the additional pull of its neighbouring moons. This gravitational tug of war has a miraculous effect on Io, transforming it from a dead world into one of the most dynamic geological bodies in the Solar System.

Io circles Jupiter every 1.77 days but crucially, for every orbit of Io, Ganymede goes around almost exactly four times and Europa goes around just twice. This beautiful symmetry is no coincidence – it is a consequence of the same complex gravitational dynamics that locked our moon's face forever towards Earth. The technical term for this relationship is 'orbital resonance'. Whilst the mathematics and terminology might be complicated, the effect is simple to explain. Periodically, Io, Europa and Ganymede line up together, and when they do Io gets a powerful gravitational kick on a regular basis. This has the effect of forcing Io out of a nice circular orbit into an elliptical or eccentric one. This means

that Io moves closer and farther away from Jupiter in every orbit as it orbits in an ellipse with the gas giant at one focus. Because Jupiter's gravitational field is so great, that has the effect of continually stretching and squashing Io as it sails periodically closer to and farther from the planet. This is exactly the same effect that creates Earth's ocean tides, but on Io it is not water that is pulled and pushed, but the solid rock of the moon itself.

Just like a squash ball, as Io is stretched and squashed repeatedly it heats up by the friction, transferring vast amounts of energy from its orbit to the rocky interior of the moon itself. Over time, this vast energy transfer would cause Io's orbit to become more circular, but the elegant relationship between its orbit and its sister moons Europa and Ganymede ensures that it must continue its eccentric elliptical path around Jupiter. This relentless gravitational tug of war keeps Io literally boiling hot, moving rock as if it were nothing more than water and transforming it into a world seething with heat, alive with volcanic activity. As the images came back from Voyager, it was immediately clear that the three scientists were right: Io was anything but a dead dusty moon.

We have since sent a probe even closer to Io. At the end of the last century, the Galileo spacecraft took the best images we have to date of Io, helping to create a detailed picture of the geological life of the volcanic bubbling moon. We now know that just one of the many lava lakes on Io releases more heat than all Earth's volcanoes put together. The lava lakes on Io are vast; the largest is 180 kilometres (112 miles) in diameter, dwarfing magnificent Erta Ale. Io's surface is covered with hundreds such volcanic centres, making it by far the most volcanic place in the Solar System, endlessly pumping out heat into the cold vacuum of space.

Io is a surprising and bizarre world. Being so far from the Sun, its surface is about -155 degrees Celsius. It is covered in frozen sulphur, giving it a rich yellow colour. Yet amidst the frigid yellow planes, Io is scarred by cauldrons of molten lava; thousands of tons of flowing rock melted by energy extracted from Jupiter's powerful gravitational field as it traces its eccentric orbital path ◉

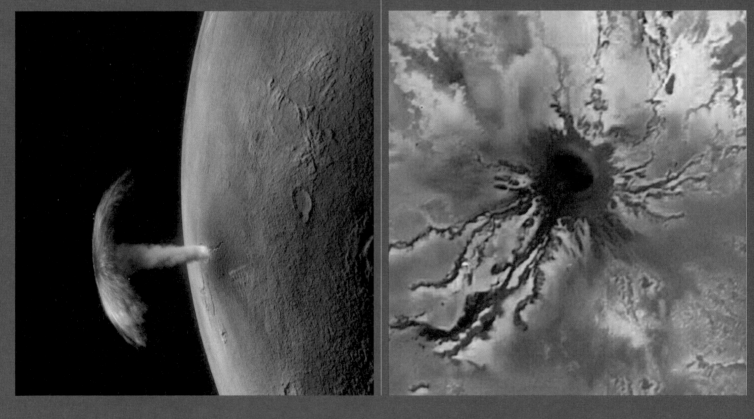

VOLCANISM ACROSS THE SOLAR SYSTEM

Volcanism is found throughout the Solar System. It is generally produced by either internal heat, as found on Venus, Earth and Mars, or by tidal heating caused by gravitational pull, as found on Io and Enceladus.

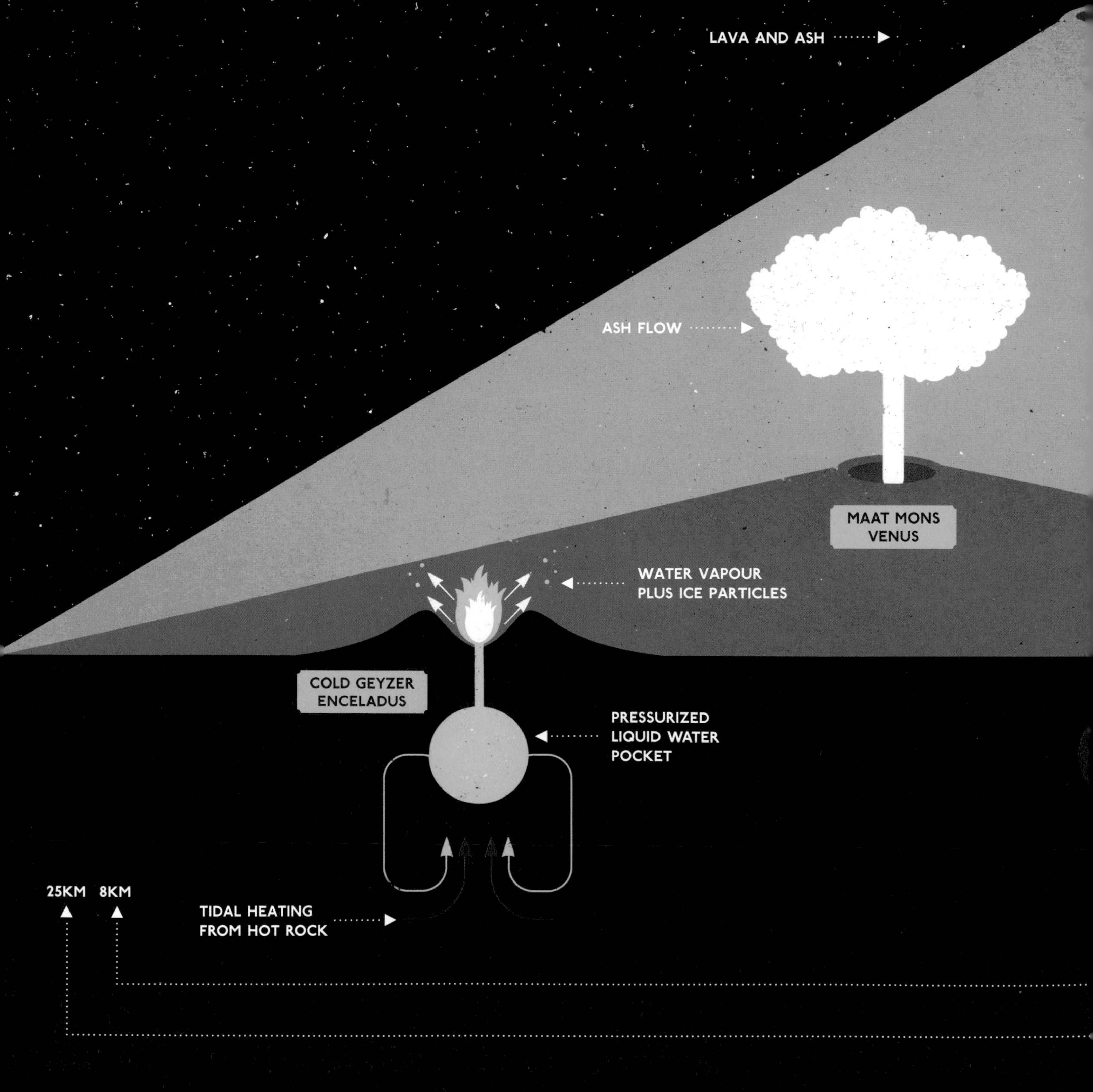

LAVA AND ASH ┈┈┈┈⟶

ASH FLOW ┈┈┈┈▶

MAAT MONS
VENUS

WATER VAPOUR
PLUS ICE PARTICLES

COLD GEYZER
ENCELADUS

PRESSURIZED
LIQUID WATER
POCKET

25KM   8KM

TIDAL HEATING
FROM HOT ROCK ┈┈┈┈▶

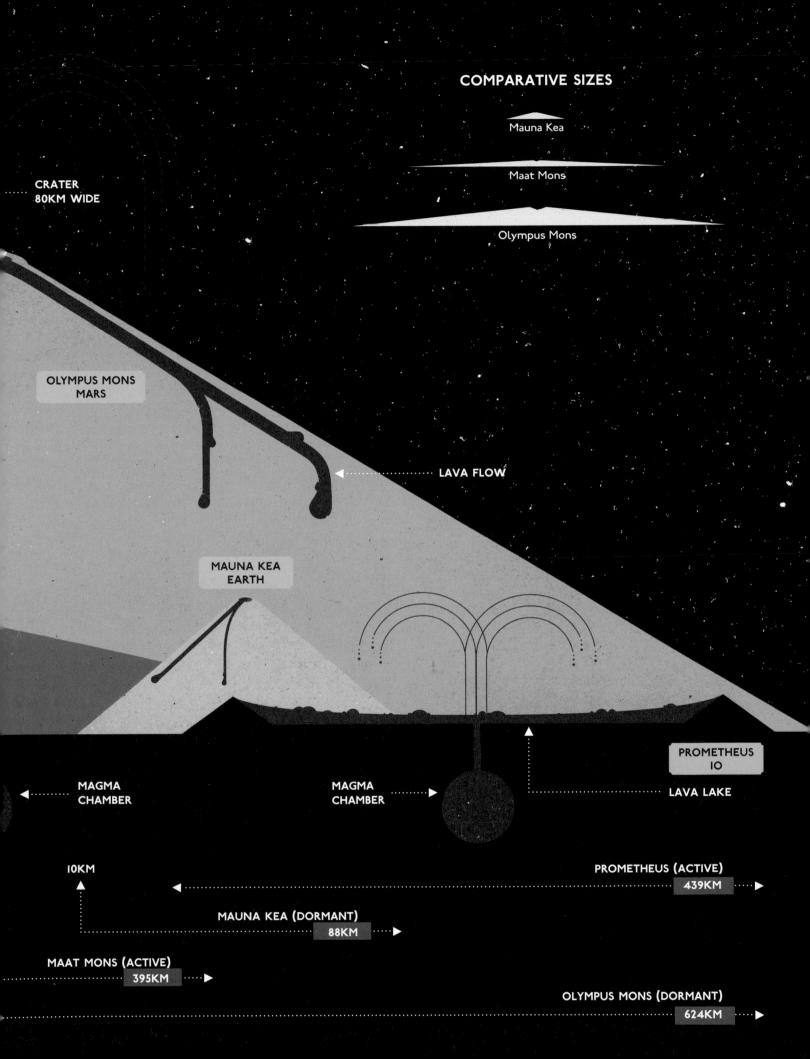

COMPARATIVE SIZES

Mauna Kea

Maat Mons

Olympus Mons

CRATER
80KM WIDE

OLYMPUS MONS
MARS

LAVA FLOW

MAUNA KEA
EARTH

PROMETHEUS
IO

MAGMA
CHAMBER

MAGMA
CHAMBER

LAVA LAKE

10KM

PROMETHEUS (ACTIVE)
439KM

MAUNA KEA (DORMANT)
88KM

MAAT MONS (ACTIVE)
395KM

OLYMPUS MONS (DORMANT)
624KM

THE ECHOES OF THE SOLAR SYSTEM

Io is a world beyond our imagination. Its unique gravitational connection to its parent planet provides a seemingly inexhaustible supply of heat. As well as its huge lava lakes, the heat also powers the largest volcanic eruptions in the Solar System. Molten rock and gas blast out from the frigid surface; the gas expands, shattering lava into giant fountains of fine particles. With weak gravity and a sparse atmosphere, Io's volcanic plumes tower over 300 kilometres (186 miles) above the moon's surface.

This incredible phenomenon, volcanism, comes from the simplest of laws of physics; the heat contained within a planet will eventually find a way to escape into the coldness of space. But what a spectacular way for the laws of physics to play out!

In the most unexpected of places, in the coldest reaches of the Solar System, the simple flow of heat has created a fiery world of wonder. And as we have seen, Io is not alone. We have discovered that many of the moons in the Solar System are far from dead, barren and uninteresting worlds; they are active, sometimes violent and always beautiful.

Io is fascinating. It doesn't derive its energy from an internal heart source in the same way that the Earth does; it extracts energy from its orbit around its parent planet, Jupiter. Having lived for three nights beside the magnificent, brooding pretence of Erta Ale, I can barely begin to imagine what an astonishing sight the vast lava lakes on Io must be. Io is indeed a true wonder of the Solar System.

Our exploration of the planets and moons orbiting our star has given us valuable insights into the nature of our own world, and changed our view of our planet's place in space. Out there are many truly violent and hostile worlds, but they are driven by the same laws that shape and control our own world. The laws of Nature can create vastly different worlds, given the tiniest of changes in temperature and composition. Worlds can also be profoundly changed by the influence of neighbouring planets and moons. Their very life and death is governed by delicate gravitational interconnections that span the Solar System. In fact, we might not be here if it weren't for these subtle connections.

Perhaps the most profound lesson of all is that we don't live on a planet isolated from the rest of the Solar System; there are echoes of other planets on Earth. We live in a place that is intimately connected to our sister worlds, orbiting around the star we all share ◉

CHAPTER 6

ALIENS

LIFE ON EARTH

I think we're living through the greatest age of discovery our civilisation has known. We've voyaged to the farthest reaches of the Solar System; we've photographed strange new worlds, stood in unfamiliar landscapes and tasted alien air. The one thing we haven't found on those worlds is the thing that makes our planet unique: life.

But is that really true? Is the Earth the only place in the Solar System that could support life, or are there other worlds that also harbour the conditions to do so? What we find on these worlds may help us to answer the question: Are we alone in the Universe? It's not only one of the great fundamental questions for science, it's also one of the great unanswered questions in human history.

BELOW: The research vessel Atlantis cruises the Sea of Cortez, Mexico. It is the mother ship for the submarine Alvin, which is built like a spacecraft for exploring the deepest depths of the ocean.

The Sea of Cortez, off the coast of Mexico, is one of the most diverse ecosystems on the planet. Visitors to this narrow strip of water include manta rays, leatherback sea turtles and many species of whales, particularly the world's largest animal – the blue whale. All of these animals, and thousands of others, combine to make it one of the most unique locations on Earth. It is the perfect place to explore the one characteristic that defines our planet more than any other. There are so many rich and diverse forms of life on Earth that, amongst the millions of species that flourish here, our interest is often only drawn to the grandest of animals. But some of the most remarkable and interesting wonders of life on Earth are hidden much further from our view.

One visitor that regularly makes the migration back to the Sea of Cortez is the research vessel Atlantis. This 92-metre (300-foot) floating laboratory is operated by the Woods Hole Oceanographic Institution and on board is a legendary intrepid explorer of our deep oceans. Alvin is a seventeen-tonne, deep ocean submersible. Built like a spacecraft, it is designed to take three lucky humans on a nine-hour journey 4,600 metres (15,000 feet) below the ocean waves. It is one of the world's most rugged submarines, and since its launch in 1964, it has explored some of the most extreme environments on Earth, including the wreck of the Titanic. Alvin is an explorer of some of the most alien environments we know.

On the morning of our Alvin dive, I confess to having been irrationally apprehensive. Irrational, because Alvin has a perfect safety record stretching back almost fifty years; apprehensive, because on the ocean floor beneath the Sea of Cortez, this little 4.9-centimetre (1.8-inch) -thick titanium sphere will be subjected to a pressure 200 times Earth's atmospheric pressure and will be utterly isolated from the

LIFE UNDER PRESSURE

At 900 m below the surface of the sea, the pressure is the same as found at the surface of Venus. Alvin has discovered life that lives under pressures that are more than twice that found on the surface of Venus.

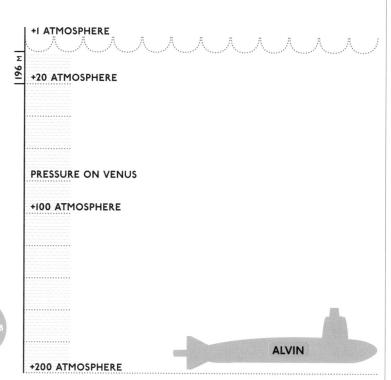

+1 ATMOSPHERE

196 M

+20 ATMOSPHERE

PRESSURE ON VENUS

+100 ATMOSPHERE

ALVIN

+200 ATMOSPHERE

BELOW: Alvin, named after the engineer Allyn Vine who was instrumental in its development, has been in operation since 1964. Deployed from an A-frame gantry on the tender ship *R/V Atlantis* (bottom), it can then submerge to over 4,267 metres (14,000 feet). It completes nearly 200 dives per year and since its first launch has found over 300 species new to science.

BELOW RIGHT: 2,000 metres (6,600 feet) below the ocean's surface, these tube worms thrive in the most extreme living conditions. From the white tubes these animals extend feathery red plumes that take in chemicals and release waste. The colonies of symbiotic bacteria that live inside the worm then convert these chemicals into nutrients on which the worm will feed.

rest of the world. It takes a space shuttle around one hour from the firing of its retro rockets in orbit to return safely to Earth; it takes Alvin two hours to return from 2 kilometres (1.2 miles) below the surface of the ocean.

Alvin is not large or luxurious; its living quarters are 208 centimetres (81 inches) in diameter, which is just big enough to allow three people to sprawl inside with legs partially intertwined – unless you can sit cross-legged for eight hours, I can't. The curved polished-titanium sides of the sphere are exposed where racks of equipment and oxygen cylinders do not obscure them – Alvin carries enough air for a three-day stay under the ocean, should rescue become necessary. The most exciting features of the vessel are three thick portholes that become beautifully and unnervingly transparent once submerged. Through these windows, generations of undersea explorers have gazed out across the ocean's most exotic and alien vistas.

Launch would be unpleasant were it not for the adrenalin of the virgin aquanaut. Alvin is swung out over a choppy Sea of Cortez on a crane and dumped into the waves, where it bobs, uncontrolled, until the final pre-dive checks are completed. The sea conditions on our dive are marginal, which makes the experience rather like I would imagine it would be sitting in a detergent ball in a washing machine. I am told it will be much worse when we re-surface, because in these conditions recovery can take up to an hour.

Within moments of being cleared to dive, however, the Alvin becomes a serene and calm place to be as it begins its long journey to the seabed. Through the portholes, darkness rapidly descends, and the only sound is the hum of the air conditioning and the occasional beep from the electronics. All sounds, including speech, have an unnatural lack of reverberation inside Alvin. It's like those first few seconds outside your house after a deep snowfall, when the world loses its echoes along with its colours. 'Beeping is never good', I say to the pilot. 'It's my pre-set depth alarms', he replies calmly.

After an hour of gentle descent, illuminated only by the magnificent flicker of bioluminescent organisms drifting past the portholes, we arrive at the ocean floor. Alvin's lights are switched on and a new world appears.

Hidden 2,000 metres (6,600 feet) below the surface of the ocean is one of the most bizarre environments on our planet. Clustered around a hydrothermal vent – a volcanic opening in the Earth's crust through which clouds of sulphurous chemicals pour into the ocean, suspended in water heated to nearly 300 degrees Celsius – is an underwater city. This miniature skyline, with fantastically complex spires reaching only a few metres into the blackness, but with an intricacy that tricks the eye and removes the sense of scale, is created by energy released through the ever-moving San Andreas Fault. Above the surface this fault is connected with death and destruction (most famously in the Great 1906 San Francisco Earthquake that devastated the city), however, below the waves the fault does not take life, but creates it.

The vast majority of the known life forms on our planet rely on energy from the Sun to fuel their existence, but deep down on the ocean floor there is no sunlight to power them. Some of the Sun's energy does make it slowly down in the form of decaying plant and animal debris from the higher levels of the ocean. This biological material has captured the Sun's energy via photosynthesis, and so delivers a slither of solar power to the ocean floor. Yet it cannot meet the energy needs of the vast density of life we see living in the dark cold depths of the ocean.

Spanning the floor of the city around the vents are carpets of yellow bacteria, forming the foundation of the ecosystem that flourishes here. Tiny shrimp-like creatures called amphipods feed directly on the bacteria. Larger organisms visit the city from the shallows, creating a complex food chain that supports a web of animals from snails and crabs to tube worms and octopuses.

Tube worms play a particularly important role in the ecology of these environments. These strange creatures can grow up to 2.5 metres (8 feet) long and spend their entire lives miles beneath the ocean. They have a well-developed nervous system, and their circulatory system uses complex haemoglobin molecules that are similar to those found in our blood to transport oxygen around their bodies. This creates the striking red plume that extends from the tip of the worms to their base, but for all their vascular complexity these animals have no mouth or digestive tract. Instead they absorb nutrients directly into their tissues through a symbiotic relationship with the bacteria that live inside them. Over half the body weight of a tube worm is bacteria, and this biological marriage is consummated by the exchange of molecules that are essential for each of these organisms to survive.

It takes a space shuttle around one hour from the firing of its retro rockets in orbit to return safely to Earth; it takes Alvin two hours to return from two kilometres below the surface of the ocean.

209

It is a relationship dependent on a chemical that is abundant around all hydrothermal vents and is essential to allow this ecosystem to survive. Hydrogen sulphide, which smells of rotten eggs, is produced when seawater comes into contact with sulphate in the rocks below the ocean floor. The bacteria living around the vents have evolved to use this molecule instead of sunlight as their energy source in a process known as chemosynthesis. Reacting hydrogen sulphide with carbon dioxide and oxygen, these unique bacteria create organic molecules that all the other organisms around them can feed off. They also produce solid globules of sulphur that give the ocean floor its vivid yellow colour. The oxygen, carbon dioxide and hydrogen sulphide are delivered to the bacteria by way of the tube worms' extraordinary circulatory systems. The red plume filters these chemicals from the seawater, then the blood transports them to the mass of bacteria in the worms' bodies. The bacteria then provide the organic compounds; the food for the worms.

This extraordinary relationship reveals the sheer adaptability of life as it evolves in the most unlikely of environments. Without the one vital ingredient associated with almost all life on the surface – sunlight – these biological renegades have found a completely novel way of bioforming.

The fascinating thing about finding life in this alien environment is that the conditions on the deep ocean floor are more similar in many ways to the conditions on worlds hundreds of millions of kilometres away in the Solar System than they are to the conditions just two kilometres above on the Earth's surface. It is incredibly dark; there is no sunlight and a brutal mixture of hot and cold water is in contact with rock and minerals. If life can not only survive but even flourish in these conditions, then it is not unreasonable to speculate that life might also survive and flourish out there in the Solar System if similar conditions are present. And as we have found again and again throughout our scientific journey to the edges of the Solar System, the way to search for the conditions necessary for life on other worlds is to actually go there and explore.

The search for extraterrestrial life stretches back thousands of years and has played a central role in both Eastern and Western thought. The Greek philosopher Thales is believed to be the first Western philosopher to introduce the idea of life outside our planet. He suggested that the stars were not just lights in the sky but other worlds, opening up the possibility of other life existing within them. Jewish, Hindu and Islamic thinkers also hinted at the idea of other worlds in their ancient literature. With the spread of Christianity in the West and its central tenet of a geocentric universe, however, speculation about the existence of extraterrestrial life became unfashionable and even opposed to Christian doctrine for many centuries.

It wasn't until the invention of the telescope 400 years ago that astronomers were able to start looking for direct signs of life on our neighbouring worlds. As technology has improved, we have been able to search the planets in more and more detail, but an increase in acuity has not always led to an increase in accuracy.

William Herschel, the discoverer of Uranus and one of the forefathers of modern astronomy, believed that every planet was inhabited; he even produced calculations to prove that the Sun was populated by a race of giant-headed beings

who could only survive in such conditions because of the size of their heads. At the turn of the twentieth century, Mars became the centre of our attention in the search for direct evidence of cosmic neighbours. American astronomer Percival Lowell convinced much of the world that Mars was covered in intricately constructed canals that were direct evidence of a complex Martian civilisation. His romantic vision of a civilisation channelling scarce Martian water from the poles to the great cities at the equator wasn't entirely dismissed until Mariner 4 arrived for a close-up look in 1965.

As our exploratory ambitions and technical prowess have increased throughout the twentieth century, the evidence for any form of living companions in the Universe has diminished. Again and again we've searched more and more carefully and found nothing. This doesn't mean the rest of the Solar System is dead; we have, after all, only scratched the surface of what's beyond us. There are literally hundreds of other worlds out there – a vast and diverse collection of planets and their moons of the Solar System that we have barely explored. Amongst them may be worlds that hold the conditions to support life, and the most accessible way to understand what the limits of those conditions might be is to look at the one place where we know life flourishes: the Earth ◉

WHAT IS LIFE?

Life is a difficult thing to define. Scientists still struggle to come up with a description that is specific enough to cover all the life forms that we already know of and broad enough to encompass the new life forms that we continue to discover here, and perhaps one day on other worlds. Many definitions have been put forward over the years that encompass the essential elements of life – metabolism, reproduction, growth, adaptation and organisation. One of the simplest definitions, the idea that life is 'a self-sustained chemical system capable of undergoing Darwinian evolution', is perhaps one of the most accurate.

Reduced to its most basic building blocks, life is nothing more than chemistry. It is an emergent property; a consequence of the many and varied reactions between a wonderfully intricate and ordered system of both simple and complex molecules. Given this, for life to exist you only need three things. First, you need the right chemistry set. A human is made up of approximately forty elements, almost half of the known elements. This is complicated, and many of them are essential to our biological function, although 96 per cent of our bodies are made of only four: carbon, nitrogen, oxygen and hydrogen. Secondly, you need an energy source; a battery that generates a flow of electrons to power the processes of life. Here on Earth, most of the life we know of uses the power of the Sun, but as we've seen around the hydrothermal vents on the ocean floor, this is not essential. As long as energy can be harvested – whether through photosynthesis capturing the power of the Sun

OUR HUMAN MAKE-UP
Around 96 per cent of our human bodies are made up of just four elements; the other 4 per cent is made up of around 36 other elements.

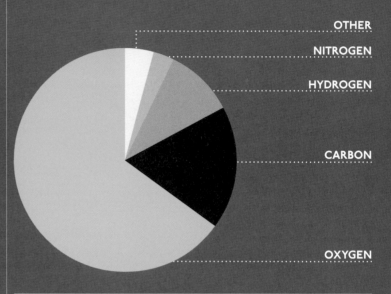

- OTHER
- NITROGEN
- HYDROGEN
- CARBON
- OXYGEN

or the chemosynthesis of hydrogen sulphide liberating the binding energy stored within molecules – life can potentially flourish. Thirdly, and seemingly universally, you need a medium through which the chemical processes of life can play themselves out. On Earth, you don't have to look far to find that medium; the solvent of life is everywhere, because it's water.

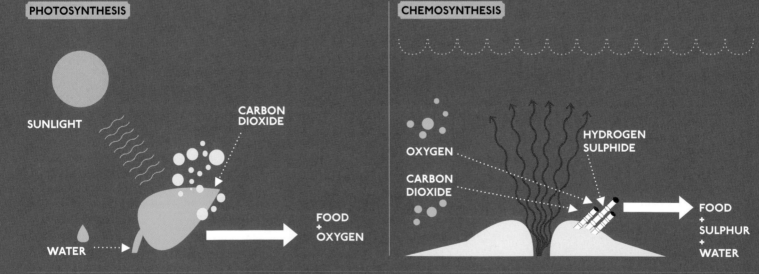

PHOTOSYNTHESIS

SUNLIGHT

CARBON DIOXIDE

WATER

FOOD + OXYGEN

CHEMOSYNTHESIS

OXYGEN

CARBON DIOXIDE

HYDROGEN SULPHIDE

FOOD + SULPHUR + WATER

HOW LIVING OBJECTS CREATE ENERGY FOR LIFE
On Earth's surface, most life forms use the power of the Sun to create food and oxygen through photosynthesis. But where sunlight is absent, such as in the depths of the dark oceans, they have adapted to use other means to create essential elements for life, such as chemosynthesis.

WATER:
AN ESSENTIAL
LIFE FORCE

BELOW: The Atacama Desert in Chile is considered the driest place in the world. In its almost complete absence of rain, no life forms can be sustained here – not even bacteria.

If you want to see how important water is to life, there's no better place to come than the Atacama Desert in Chile. The Atacama is widely considered to be the driest desert in the world. This 1,000-kilometre (600-mile) -long rainless plateau in northern Chile is sandwiched between the Andes and the Chilean coastal range of mountains. This unusual geological position creates a rain shadow, a meteorological phenomenon that prevents this thin strip of land receiving even the smallest amounts of precipitation.

All deserts are characterised by a lack of moisture, but the Atacama takes that to the extreme. Some weather stations here have never received rain; others have measured just one millimetre of rainfall in ten years. There are river valleys that have been dry for 120,000 years and there are rocks that haven't seen rainfall for twenty million years. Evidence suggests that there was no significant rainfall for 400 years in the Atacama from 1570 to 1971. It is so dry, it makes the

Sahara Desert look wet – even the great African wilderness receives fifty times more rainfall than the Atacama.

Scientists have searched for bacteria, the most basic form of life, in the Atacama and in some places they have found absolutely nothing. The land is so dead that in places the soil is more sterile than a hospital operating theatre. It's the starkest evidence we have to suggest that every life form, even the most primitive, needs water to survive. On Earth, we have found no exceptions to this rule.

This seemingly fundamental link between water and life is driving the search for life in the Solar System, because wherever we find water the evidence shows that these will be the best places to look for life beyond Earth. We are certain that Earth is currently the only planet that has standing liquid water on its surface. The other planets are either too close to the Sun and too hot – like Venus, where any water evaporated long ago – or too far away – like Mars, where the only surface water we know of is locked away in the polar caps. Further out in the Solar System, there is plenty of water. Many of the moons around the gas giants and even the rings of Saturn are made of large quantities of water, but in the depths of space it is frozen into solid ice.

However, this emphatically does not signal the end of the search. Water may be hiding below the surface of a moon or planet, and it is also very possible that liquid water once flowed across the surfaces of other planets at some point in the Solar System's history. If it did, we should be able to find the evidence, because the one thing our study of the Earth's landscape tells us is that wherever water goes, it always leaves its footprint ◉

THE SIGNATURE
OF WATER

ALL IMAGES: For the ultimate demonstration in how water sculpts our landscape, there is no better place to visit than the spectacular Scablands in the northwestern United States.

Nowhere on Earth is the faded presence of once-flowing water more obviously imprinted on the land than in the extraordinary landscape found in a remote part of the northwestern United States. The Scablands are a unique geological formation that stretch across a vast area of the state of Washington, demonstrating on a spectacular scale how water can carve its signature into rock.

First studied in the 1920s, the origin of the Scablands was a mystery that defied any conventional geological explanation. A normal river valley leaves behind a characteristic V-shaped cross section that is replicated in river systems across the globe. The other commonly observed effect of water are the U-shaped valleys carved by glaciers. The valleys of the Scablands have a rectangular cross section, however, and when first studied this could not be explained by any known geological phenomenon. It is obviously not a normal river system, because all the valleys are carved in straight lines through the rock. There is no gentle meandering of a river; instead these valleys simply look like great big rectangular holes.

J. Harlen Bretz was the first geologist to study this area. He concluded that the unique geometric erosion patterns, potholes and ripple marks of the Scablands were caused by a vast quantity of water, hundreds of cubic kilometres in

volume, passing though the area in a very short period of time. Without any explanation for where the water came from, his theory was ridiculed and dismissed. It was only when he began collaborating with another geologist, J. T. Pardee, that they were able to offer a full explanation for this extraordinary site. After decades of careful research, Bretz and Pardee arrived at a theory that not only explained the observed features of the Scablands, but subsequently helped geologists to understand similar formations on another planet millions of kilometres away.

Today, the Scablands are an area of intense interest for astro-geologists such as Jim Rice from the Arizona State University. Rice believes that understanding the events that created this landscape can help in the search for water on other planets. As our helicopter flew through the geometric canyons of a landscape unlike anything I've ever seen on Earth, Rice explained to me that investigating the origin of places like the Scablands feels more like the study of a crime scene than a geological expedition. All this detective work has revealed that these unique scars in the landscape bare witness to the largest flood the Earth has ever seen.

Between 13,000 and 15,000 years ago, at the end of the last ice age, a huge glacial lake known as Lake Missoula lay 320 kilometres (200 miles) to the east of the Scablands. It

> ## 'If you took every river in the world, put them in the same location, had them flowing at the same time, these floods were ten times larger than that.'

was held in place by a vast wall of ice – a dam that held back millions of cubic kilometres of water for thousands of years during the ice age. As the water levels increased behind the dam, the ice wall was put under more and more strain until eventually, inevitably, it failed. When it ruptured, over 2,000 cubic kilometres (480 cubic miles) of water swept out in a single catastrophic event. The floodwaters were at least a kilometre deep, travelling at 130 kilometres (80 miles) per hour. The energy released was equivalent to 4,500 megatonnes of TNT. Imagine what a terrifying sight this massive wave must have been, perhaps the largest and most devastating wave in history, rumbling across the landscape loaded with huge chunks of ice from the dam and vast boulders of basal rock ripped from the ground.

As the floodwaters tore across the landscape, they carved out a thirty-kilometre (twenty-mile) -long canyon, and at its head they left giant horseshoes. At over 122 metres (400 feet) high and 8 kilometres (5 miles) across, this was the largest waterfall the world has ever known. As we stood looking across this landscape, it was virtually impossible to imagine the scale of the wave of water that had flowed through here. Jim Rice, though, had a simple way of putting the event into perspective: 'If you took every river in the world, put them in the same location, had them flowing at the same time, these floods were ten times larger than that.'

What's perhaps even more astounding, though, is the speed with which this landscape was formed. Current estimates suggest that this vast and complex landscape was created in no more than a week, perhaps even as little as forty-eight hours. As Jim Rice put it: 'It's instantaneous geology' on an epic scale.

The Scablands reveal one of the characteristic signatures that water can carve into the landscape. It is a signature so bold and vivid that it can even be seen from Earth's orbit, and if we can see these features on our own planet, then when we turn our telescopes outwards we must therefore be able to look for similar evidence of water's mighty work on the surface of other worlds ◉

217

LIFE ON MARS?

For over a century, Mars has been held up as a prime candidate for a location on which alien life might be found. For a brief moment the astronomical imagination of Percival Lowell convinced us of the existence of a network of canals on the red planet, and popular culture became suffused with tales of Martians, such as in H. G. Wells' science fiction classic, *The War of the Worlds*. The enduring legacy and romance of the Victorians' imaginary Martians persisted throughout the twentieth century, and in many ways it is still with us. However, today the evidence we are looking for is not the handy work of extraterrestrial builders, although just like Lowell we are still searching diligently for the tell-tale signs of waterways on the red planet.

OPPOSITE: This image was taken by Mars Exploration rover Opportunity near the rim of the Victoria Crater, an impact crater about 800 metres (2,600 feet) in diameter near the equator of Mars.

SCARS ON MARS

BELOW LEFT: The horseshoe shapes present at the dry falls in the Scablands (bottom) are also present in this image of Mars' surface (top).

BELOW: These two images (Scablands bottom, Mars top) demonstrate again the similar characteristics shared by both landscapes.

Images taken by the Mars Exploration rover Opportunity, which landed on the red planet in 2004, reveal that our next-door neighbour has features cut into its surface that are almost identical to those in the Scablands. Mars is covered in outflow channels – straight wide canyons exactly like those of Washington – filled with identical geological features. This all suggests that similar huge floods might have torn across the surface of the planet.

The pictures here reveal just how similar the landscapes appear to be. Viewed from the air, we can see that the horseshoe shapes of the dry falls in Washington state are replicated on the surface of Mars. And there are other striking similarities. Upstream of the falls on Earth and Mars, we can see grooves cut into the landscape as the water cascaded down and flowed over the falls. This collection of similar geological characteristics suggests that the same story played out in these gigantic valleys, and that vast amounts of water rapidly flowed over the surface of Mars at some point in the past.

These images and their interpretation, combined with the many other geological features we have studied on Mars, make a compelling case that very large volumes of liquid water once flowed across the surface of the red planet. This is an important step in the search for Martian life, because it shows us that one of the non-negotiable prerequisites for life – water – must have been present. But on their own, Scabland-like

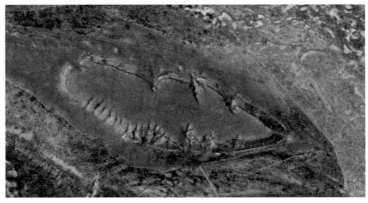

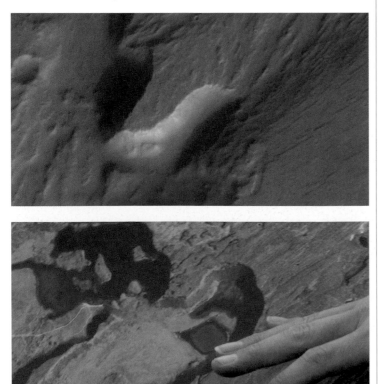

These images ... combined with the many other geological features we have studied on Mars, make a compelling case that very large volumes of liquid water once flowed across the surface of the planet.

erosion features do not point to the existence of the conditions that we believe life requires. If the same processes that formed the Scablands on Earth formed the Martian landscapes, the floods that created them may only have lasted a matter of days, and for life to get a foothold you need more than that; you need areas of standing water, lakes and rivers that persist for many millions of years. In order to look for evidence of that standing water, we've done the only thing that we can. We've sent an army of robotic explorers to the surface of the planet ◉

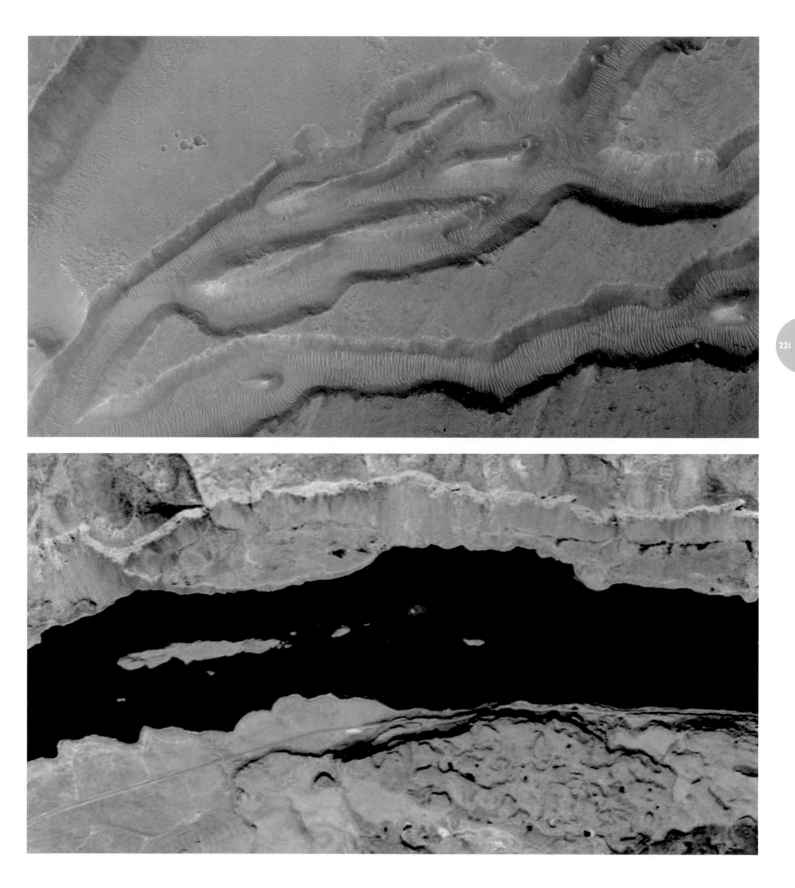

BELOW: These outflow channels that cover the surface of Mars appear to have exactly the same formation as those in the Scablands and are filled with identical geological features.

221

MARS' MINERALS

Over the last thirty-five years, we've landed six robot probes on Mars, and one of them, Opportunity, is still rolling across the surface in 2010, investigating today's Martian geology. Opportunity and her sister craft, Spirit, have captured many people's imaginations, not least amongst the school-age space enthusiasts who will form the next, vitally important, generation of explorers, scientists and engineers. The rovers genuinely are explorers in the old-fashioned sense; they are the direct extension of our senses to the surface of another world. They have also been of priceless scientific importance because you can't really get to know another planet from orbit. You've got to get down to the surface; you've got to touch it, and you've got to dig down and examine it microscopically. By doing just that, the rovers have made some extremely important scientific discoveries.

One of the most significant of those discoveries was made in November 2004. The Opportunity rover was examining an impact feature called the Endurance Crater,

LEFT: The world's largest salt works are on the Baja peninsula, Mexico. There, lagoons are pumped full of seawater, which then evaporates, leaving behind its precious residue: salt.

BELOW AND BOTTOM: These images from the rover Opportunity show just some of the gypsum crystals found in the Endurance Crater and networks of sand dunes on Mars, which suggest that large areas of the planet were once covered in water.

when it detected deposits of a remarkable mineral: gypsum. This is a very soft mineral that has been discovered in large deposits on the surface of Mars, and although we can't yet bring any Martian gypsum mineral back to Earth to test it further, its very existence provides another crucial piece of evidence in our search for life on Mars.

On the Baja peninsula in Mexico, the largest salt works in the world stretch across a vast landscape. It's a lucrative but simple industry that requires seawater to be pumped into lagoons where it evaporates, leaving behind a residue of sodium chloride, or table salt, that eventually finds its way to millions of dinner tables across the world. It's not just table salt that appears from the seawater; other salts and minerals crystallise out and emerge at different stages of the process. In one of the lagoons, pond number 9, the seawater is at exactly the right concentration to precipitate out a very beautiful crystal that covers the entire floor of the lagoon.

These crystals are gypsum; exactly the same stuff that the Opportunity rover found on the surface of Mars. What's interesting about Opportunity's discovery is that it tells us something fundamental about the story of water on the red planet. The chemical formula of gypsum is $CaSO_4$ $2H_2O$ – calcium sulphate dihydrate. The dihydrate is the bit that's important for our story, because that refers to the two molecules of water bonded loosely to the calcium sulphate. We know of only one way in which calcium sulphate can combine with water like this here on Earth; it simply requires calcium and sulphate ions to be in the presence of liquid water that is stationary for long periods of time.

With large deposits of gypsum found at multiple locations across the surface of Mars, it seems the only conclusion to draw is that there must have been standing water on the planet's surface at some point. If this interpretation of the discovery of gypsum is correct, then it is another crucial piece of evidence to suggest that all the factors needed for life to occur have existed at some point in Martian history.

Everywhere we look on Mars, from the microscopic evidence left in the rocks to the subsequent discoveries of gypsum in amongst the networks of sand dunes and the vast geological structures that are so suggestive of flowing water that cover the planet's surface, it is difficult to escape the conclusion that Mars was once a much warmer and wetter planet. A planet with oceans and floods, vast areas of standing water and a hydrological cycle that created the ghosts of a familiar Earth-like landscape that we glimpse today through the arid dust.

Although Mars may once have been a more hospitable place, any liquid water has long since disappeared from its surface. About three billion years ago, Mars died as a planet. Its core froze and the volcanoes that had produced its atmosphere seized up. The solar winds then stripped away the remains of that atmosphere. Any liquid water left on the planet's surface would have evaporated or soaked into the soil, where it froze. This left the surface of Mars too cold, too exposed and too dry to support life, but that is not to say that life couldn't exist somewhere on the red planet today. Maybe we're just looking in the wrong place; maybe there are other potential habitats for life on Mars ◉

Mars was once a much warmer and wetter planet. A planet with oceans and floods, vast areas of standing water and a hydrological cycle.

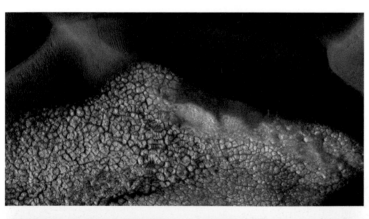

MARS' UNEXPLORED SUBTERRANEA

In September 2007, the NASA Mars Odyssey spacecraft discovered seven strange circles high up on the slopes of a Martian volcano known as Arsia Mons. These very dark circular shapes varied in size from 100 to 250 metres (330 to 820 feet) in diameter and they completely puzzled the scientists who found them. To try and solve the mystery, the team deployed the Odyssey craft's infrared cameras to record the temperature swing of the holes across a series of Martian days. The results that came back were surprising. The temperature change from day to night in these holes was much less than the change in the surrounding area; about a third of the temperature swing seen outside of the circles. Such stable temperatures that seem to iron out the change from day to night are seen all over the Earth. Caves on Earth maintain a constant temperature; the deeper they are, the more they can resist the effect of the passing Sun outside. It's why so many life forms use caves across our planet as shelter and it's why the NASA scientists realised that they were looking through seven mysterious doorways into the unexplored world of subterranean Mars.

The scientists at NASA called these holes the Seven Sisters and named them Dena, Chloe, Wendy, Annie, Abby, Nikki and Jeanne. Three images of 'Annie' taken by the spacecraft show how its opening, the size of two football pitches, is colder then the adjacent area in the afternoon and warmer than the immediate surface at night. Nobody can be sure whether these circles are deep openings into an expansive cave system or narrow vertical shafts, but what is certain is that they open up another front in our search for life on Mars.

These caves are almost certainly at too high an altitude to have supported any form of microbial life in the past or present, but their very existence opens up the possibility that somewhere on Mars there may be a cave system that could have protected life from the hostile environment outside. We know there may be water down there, too: satellite data shows permafrost, ice frozen in the soil. Deep below the surface, that ice may melt to form liquid water. These are tantalising glimpses into a hidden world that may be harbouring Martian life, concealed behind the darkness of these cave entrances. These may not appear to be the perfect conditions for life as we know it on our own world, but our exploration of life on Earth also suggests that not all living things are always so fussy ◉

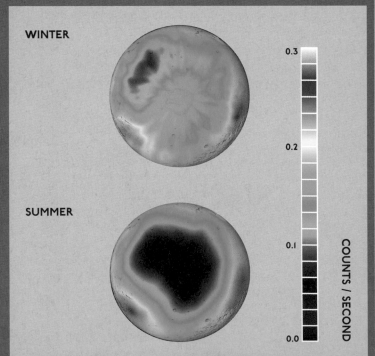

WINTER

SUMMER

0.3

0.2

0.1

0.0

COUNTS / SECOND

ABOVE: (i) Images are from Odyssey's neutron and gamma ray detectors (ii) blue colour demonstrates water-ice (iii) in winter, the water-ice is hidden by a layer of dry-ice (frozen carbon dioxide) (iv) in spring/summer, the carbon dioxide is heated and dissipates, revealing large quantities of water-ice across the north pole of the planet.

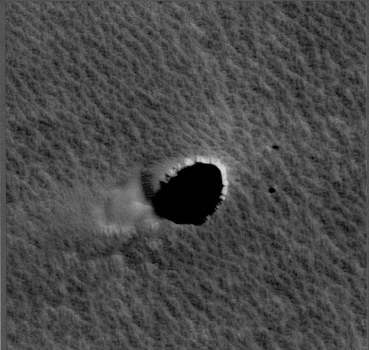

ABOVE: Tharsis Montes is the largest volcanic region on Mars. It is approximately 4,000 kilometres (2,500 miles) across, 10 kilometres (6 miles) high and contains twelve large volcanoes. The largest volcanoes in the Tharsis region are four shield volcanoes named Ascraeus Mons, Pavonis Mons, Arsia Mons and Olympus Mons.

LIFE
UNDERGROUND

The Cueva de Villa Luz in Tabasco, Mexico, the 'cave of the lighted house', is the very definition of a hostile environment to a human being.

Here on Earth, it is easy to jump to the conclusion that the perfect habitat for life would look like the green, lush landscapes around the jungle river of southern Mexico. This is the very definition of biodiversity: a warm climate, lots of liquid water, a beautiful dense atmosphere and an abundance of living things in both number and variation.

All life in the jungle, and indeed most of the life forms we meet every day and are familiar with, thrive in pretty much the same conditions that we do, driven by the heat and light of the Sun. The more sunlight and the more water present, the more life appears to like it, but just a few kilometres away from here is a form of life hidden deep beneath the surface of our planet, one that flourishes in a completely different environment and may hint at the forms of life that could be hidden away on Mars.

The Cueva de Villa Luz in Tabasco, Mexico, the 'cave of the lighted house', is the very definition of a hostile environment to a human being. This subterranean maze, over 2 kilometres (1.2 miles) in length, is full of hydrogen sulphide gas, pumped into the cavern by a spring rich in this corrosive gas. The gas dissolves in the water to produce sulphuric acid that has eaten its way through the limestone rock to create the cave system. It not only pungently smells of rotten eggs,

225

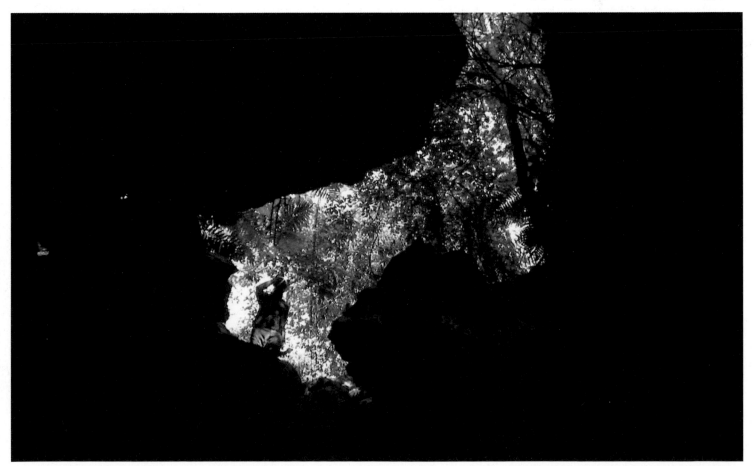

BELOW: The tiny snottites use the hydrogen sulphide gas around them to drive their metabolism, producing a secretion so acidic that it would have burnt my skin if I'd touched it while taking a test sample. It has a pH level of almost zero – making it a highly concentrated and dangerous acid.

it is also incredibly toxic to humans, even in small quantities. To step inside, you need to be kitted out with a gas monitor and a gas mask in case it all gets too much. It's a place where, at first sight, you would not expect a great many life forms to survive and flourish. Although the cave is a potential death trap for us, that doesn't mean that nothing lives here. In fact, it's teeming with life.

Everywhere in the cave water swim strange fish called *Poecilia sulphuraria*, which are beautifully adapted to tolerate these conditions. If you look at them closely, they're quite pink, and this is thought to be because they have large quantities of haemoglobin in their blood that allows them to move around using the sparse quantities of available oxygen in the water surrounding their bodies. They are a beautiful example of how life adapts to the most inhospitable of environments, adjusting the parameters of biology to suit their external environment.

Delve deeper into this toxic cave and things become even more intriguing. In the depths of the caves, where the concentration of poisonous gas sets off every alarm, is a profoundly interesting organism. Down here, far from the light of the Sun, is a life form that derives its energy not from the Sun but from the noxious gases around it. These tiny creatures use the hydrogen sulphide gas bubbling up through the springs to drive their metabolism. The very same gas that would squeeze the life out of a human is the breath of life for these creatures.

These are the snottites. They are surely one of the most alien life forms on Earth because they metabolise hydrogen sulphide. They react this nasty gas with oxygen to produce sulphuric acid, a totally exotic form of respiration. Whereas we breathe in oxygen and react it with sugars to produce carbon dioxide and energy, these little things breathe in hydrogen sulphide and oxygen and produce sulphuric acid.

It's a form of life so corrosive to us that we can't even touch it safely. The secretion dripping off the snottites has a pH of almost zero – it's a highly concentrated sulphuric acid that is as brutal as battery acid. By every definition of the word, these organisms are alien life forms, with the one discrepancy being that they are living just below the surface of our planet.

Surprisingly, the snottites are not alone. Organisms that can extract energy from the minerals around them are found under the ground all over the world. In fact, this way of life is so successful that it's thought there may be more life mass living beneath the Earth's surface than there is on it, and that raises an intriguing possibility. If life can thrive below the Earth's surface, why couldn't organisms like snottites survive and flourish beneath the surface of Mars? Living below the surface of Mars might in fact be quite a good idea, because the surface is incredibly hostile. Any life form on the Martian surface would be subjected to intense ultraviolet radiation from the Sun. Mars is also a very cold place, and the atmospheric pressure doesn't allow liquid water to exist on the surface.

If there is life below the surface of Mars, however, then obviously we would appear to have a problem detecting it. Interestingly, there is one last tantalising clue for us that suggests there might be something going on below the Martian surface ◉

THE FINAL PIECE OF
THE JIGSAW

In 1768, the Italian physicist Alexander Volta was still thirty years away from making his most famous contribution to science: the development of the first battery. As a totally unconnected preamble to his great invention, save for a demonstration of his intense curiosity about Nature, he spent his time collecting gases from the marshes around his home in the north of Italy. One of the gases he collected and studied was methane, a simple molecule with the chemical formula CH_4. Volta demonstrated with great flair that methane could be ignited from an electric spark. Today, we use this combustible gas as one of our main sources of fuel, because methane is the main component of natural gas and its abundance makes it widely available and relatively cheap. The deposits of methane under the Earth's surface are generated by the decay of organic material and are often linked to sites containing other fossil fuels, but methane is also present in our atmosphere.

Termites, or white ants, are very unusual animals because they eat dead organic matter; their primary diet is wood. There are many species of these insects, billions of individuals across the planet, and in the process of digesting wood they produce vast quantities of methane, pumping an estimated fifty million tonnes of it into the Earth's atmosphere every year.

Termites aren't the only methane producers on the planet, though. There is lots of methane occurring naturally in our atmosphere, produced either biologically or by active geological processes such as mud volcanoes. That makes it all the more surprising that methane has been detected in the atmosphere of Mars, a planet we have thought of as both geologically and biologically dead. As far as we know, there is no non-biological or geological process that could produce and sustain the levels of methane we observe in the Martian atmosphere.

In January 2009, researchers at the Keck Observatory in Hawaii announced that they had discovered substantial pockets of methane on the surface of Mars. Using the infrared telescope facility on top of this famous volcano, the NASA-led team had previously detected only tiny amounts of methane in the Martian atmosphere. Closer observations subsequently revealed that the gas was being produced in far greater quantities than had been suspected, although

LEFT AND BELOW RIGHT:
Termites, or white ants, are very
unusual animals because they feed
on decaying wood. As they eat the
wood across the globe, they produce
methane, contributing fifty million
tonnes of the gas into the Earth's
atmosphere every year.

NATURAL SOURCES OF ATMOSPHERIC METHANE
Well over half the methane production on Earth stems from human activity,
but it is also produced by natural sources in significant quantities, which
could explain methane's presence on Mars.

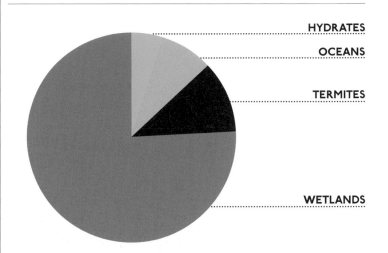

HYDRATES

OCEANS

TERMITES

WETLANDS

it is concentrated in a handful of plumes that vary with the
seasons. In the warmer summer months, thousands of tonnes
of the gas are released from vents in the surface. As far as
we know, this can mean only one of two tantalising things.
The simplest explanation is that, just as here on Earth, the
methane plumes are emanating from previously unknown
geological processes. This would be a great discovery in itself,
because it would mean that far from being dead, Mars is
still a geologically active planet. The other option is that the
methane is being produced from a biological source; living
organisms on Mars that are producing methane just like the
termites on Earth.

No one is seriously suggesting that there are termites
running around beneath the surface of Mars, but it's not
actually the termites that are particularly interesting about
this story – it's the way they digest the wood. They use
symbiotic micro-organisms called *Archaea*, which live in
their guts, to do the methane-producing work for them.

Archaea used to be thought of as an unusual group
of bacteria, but we now know that these basic living units
are in fact a completely distinct branch of life. Along with
bacteria and the more complex cellular units known as
Eukaryotes, Archaea are one of the three separate branches
of life on Earth. They are found all over our planet; filling the
soil, the oceans and the guts of humans, cows and termites.
They are also the most common organisms found beneath
the surface of the Earth.

Archaea thrive in many of the Earth's most extreme
environments. The snottites, as seen in Cueva de Villa Luz,
are members of the *Archaea*, as are many of the micro-
organisms found living around deep-sea hydrothermal vents,
producing millions of tonnes of methane that is pumped
into the atmosphere. As we learn more about the *Archaea* on
Earth, it raises the tantalising prospect that the methane we
see in Mars' atmosphere might just be produced by organisms
like these living below the Martian surface.

Going the extra vital step to proving this incredible
hypothesis for the origin of the Martian methane will almost
certainly involve a manned mission to Mars, or at the very
least the development of another advanced generation of
robotic explorers. With so much evidence calling us back
to explore, it seems impossible not to imagine that we will
take this final step in our long search for companions in the
cosmos. The idea that we are potentially just one mission
away from the most profound discovery in human history
must surely be reason enough to embark on at least one more
journey to the red planet of Lowell and Wells ◉

*The idea that we are
potentially just one mission
away from the most profound
discovery in human history
must surely be reason enough
to embark on at least one more
journey to the red planet.*

EUROPA:
LIFE IN THE FREEZER

While Mars remains one of the prime candidates in the search for extraterrestrials, it is no longer the only place in the Solar System that we think could harbour alien life. As we leave behind the familiarity of the rocky planets and travel further from the Sun, the Solar System becomes a very different place. In our search for water, the far-flung reaches of the outer solar system provide a plentiful source of H_2O, but half a billion kilometres away from the Sun, any water might be expected to be frozen as hard as steel.

OPPOSITE: At first sight, Jupiter's frozen moon, Europa, seems to harbour conditions too hostile to support any form of life.

Jupiter, the vast gas giant, is surrounded by a network of sixty-three moons, with many containing vast amounts of ice. The largest four are the moons discovered by Galileo in January 1610. Out here, where the Sun is little more than a bright star in a dark sky, you might imagine these moons to be cold, desolate places, and that's how they appear at first sight.

Callisto is the most distant of the Galilean moons from Jupiter. Orbiting not far short of 2 million kilometres (1.2 million miles) away from its mother planet, Callisto takes 16.7 days to complete an orbit and the same amount of time to turn on its axis and complete a Calliston day. A giant moon, it's the third-largest satellite in the Solar System, and analysis suggest that it is made up of approximately half water ice. In this foreboding part of the Solar System, the ice is frozen as hard as steel on its surface, at a temperature of -155 degrees Celsius.

Travelling 800,000 kilometres (500,000 miles) towards Jupiter, the next Galilean moon is the largest moon in the Solar System – Ganymede. With a diameter of over 4,800 kilometres (3,000 miles) this moon is bigger than the planet Mercury. Composed of silicate rock and ice, Ganymede has an uninviting surface temperature of -160 degrees Celsius.

Callisto and Ganymede don't seem likely places on which to find life. Despite an abundance of water, the freezing temperatures on their surfaces mean the solvent of life is locked up in kilometres of thick ice, making these moons seemingly far too hostile to support life of any kind.

Looks can be deceiving, though, and among these frozen wastes we have found one world that is of particular interest in our search for life beyond the planet Earth; a world that seems to defy its freezing position in the outer reaches of the Solar System.

BELOW: These images, taken by NASA's Galileo spacecraft in June 1996, show the cracked surface of Jupiter's moon Europa. The top image shows Europa from a distance of 5,340 kilometres (3,310 miles), which reveals crustal plates on the surface, and the bottom one from about 2 kilometres (1.2 miles), which clearly shows the parallel ridges on its surface.

Beyond Callisto and Ganymede, as we journey towards Jupiter, is the ice moon Europa. It's about the same size as our moon and is the smallest of the four Galilean satellites. It orbits Jupiter in just over 3.5 days at an average distance of 671,000 kilometres (417,000 miles) from the planet. It has a tenuous atmosphere, composed of oxygen, but this sliver of gas is not the reason for our fascination with this moon.

When we look closely at the moon's surface, it becomes truly intriguing. Europa is the smoothest body in the Solar System; its surface is made of an unbroken shell of ice at a temperature of a chilly -160 Celsius. Etched into this ice is a network of mysterious red markings. This seems an incredibly unlikely home for life, and the photographs taken by the Galileo spaceprobe seem at first sight to confirm that suspicion. Just like Callisto and Ganymede, Europa appears to be a vast icy wilderness, but if you look more closely you start to see features that hint at a very different story. Beneath its thick layer of ice, Europa holds an astonishing secret.

Images taken by the Galileo spacecraft in 1998 revealed something quite remarkable about Europa's icy surface. One image of the Conamara region (named after a district in the west of Ireland), reveals deep cracks that crisscross the surface of Europa. At higher magnification, we see even more

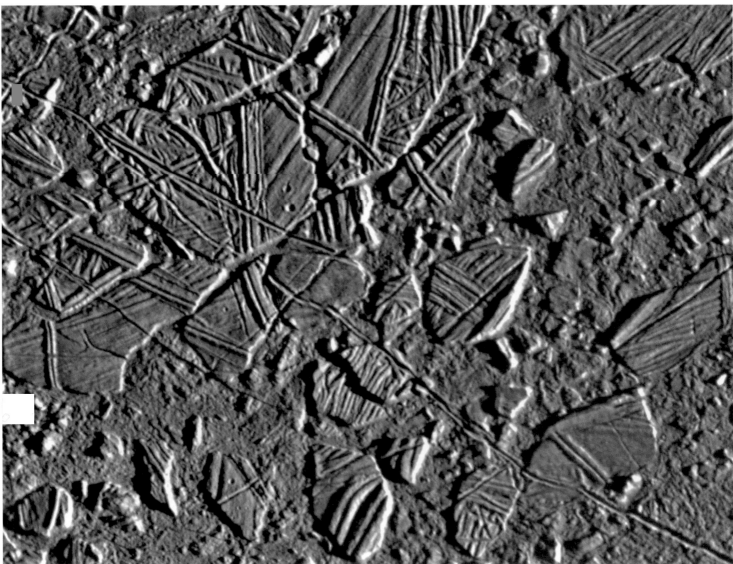

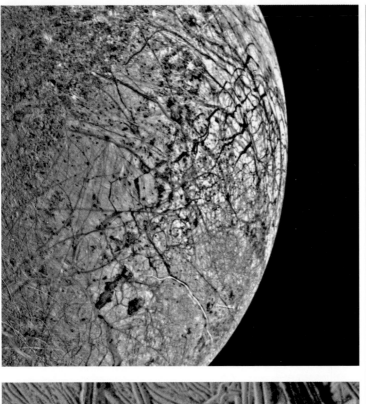

LEFT: This early image of Europa was taken by Voyager in 1979, from a distance of 241,600 kilometres (149,800 miles). The visible complex patterns suggested there were cracks over the moon's surface.

BELOW: This photograph of the Getz Ice Shelf along Antarctica's Amundsen Coast, taken by NASA scientists, reveals the similarity between Earth's icebergs and the surface of Europa.

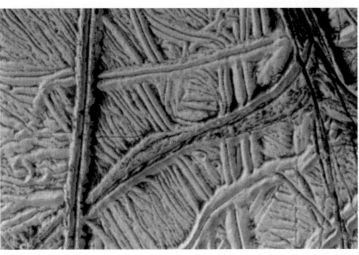

Deep cracks crisscross the surface of Europa, where ice has been broken into icebergs and jumbled up before freezing.

complexity on the surface – areas where the ice has been broken into icebergs and jumbled up before refreezing.

Compare this to an image of sea ice on Earth and the similarity is immediately apparent. These formations on our planet that resemble so closely the Conamara region of Europa are caused by the movements of the ocean under the ice that make it bend and crack. This is highly suggestive that something similar may be happening on Europa, and therefore there must be liquid water, an ocean, under Europa's icy shell.

Since the Voyager spacecraft first photographed these cracks in the late 1970s, we have been studying them to try to understand the exact forces that created them. Almost twenty years after Voyager, the Galileo probe arrived and began taking images of much higher quality that allowed detailed geological maps to be made of the surface. Studying these maps and trying to understand the origin and evolution of the cracks has provided powerful additional evidence for the theory of a subterranean ocean on Europa ◉

EUROPA'S ECCENTRIC ORBIT

As Europa orbits around Jupiter, it doesn't quite follow a circle. This slight eccentricity of the orbit is maintained by the gravitational interaction with its neighbouring moons Io and Ganymede. Just as for the volcanic moon Io, the effect of the eccentric orbit on Europa is profound. Europa is stretched and squashed as it sweeps close to and then further away from Jupiter on every orbit. This causes the interior of the moon to heat up by friction, which melts the frozen ice to produce a subterranean ocean.

The effects of the tidal stretching are not just confined to heating, however. As the surface of Europa is constantly stressed, the ice fractures and cracks, but the position of those cracks is not quite where you'd expect them to be. Europa, as with many moons in the Solar System, including our own, is tidally locked to its parent planet and so always keeps the same face pointing towards it. With this knowledge, planetary geologists can calculate exactly how and where these cracks in the ice should form, but they find that only the youngest cracks are where they are expected to be, the older cracks appear to have drifted across the surface of the moon over time. The preferred explanation for this is that the moon's surface is rotating at a different rate to its interior. The icy surface of Europa has shifted since the formation of the cracks. The only way this could have happened over a short period of time is if Europa has an ocean of liquid water surrounding the entire moon, between the rocky core and the icy shell, that allows the surface to slip around freely. The cracks could then have formed and literally slipped around the moon over time.

ORBITAL RESONANCE, FORCED ECCENTRICITY

Orbital resonance occurs when the orbital period of two satellites are related by an integer ratio, causing them to periodically align. Every time the satellites are aligned, the inner satellite S2 experiences a gravitational kick from the outer satellite SI. The effect is to pull S2 away from the planet. The net effect over a long period of time is to force S2 into an elliptical orbit. Note that, at the same time, SI feels a gravitational kick from SI, also moving into an elliptical orbit.

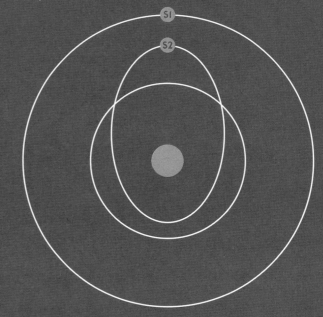

GALILEAN RESONANCE

The Galilean resonance system is made up of three satellites: Io, Europa and Ganymede in a 1:2:4 orbital resonance. For every two orbits Io makes, Europa makes one; for every two orbits Europa makes, Ganymede makes one. The satellites do not align at one common conjunction point: Io and Europa align as shown in the diagram. Europa and Ganymede have a conjunction point 180 degrees around the orbital plane.

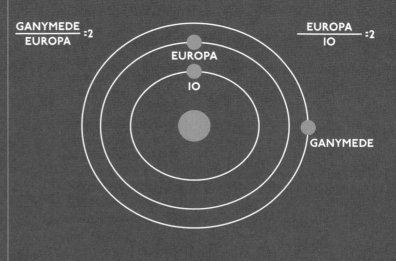

EUROPA'S INTERIOR

Beneath Europa's icy exterior is believed to lie an ocean 100 kilometres (60 miles) deep, and beneath that a rocky interior at the heart of which is a metallic core.

METALLIC CORE

ICE COVERING

ROCKY INTERIOR   **H₂O LAYER**

LIQUID WATER OR WARM CONVECTING ICE

CRACKS ON EUROPA'S SURFACE

Europa is repeatedly squeezed and stretched during its elliptical orbit, due to Jupiter's strong gravitational field. Jupiter's gravity also causes the underground ocean to be raised and lowered, just like the tides on Earth. This puts pressure on the moon's icy surface, causing it to crack. The crack is then opened and closed by tidal flexing, causing the ridge-like structures seen across the surface.

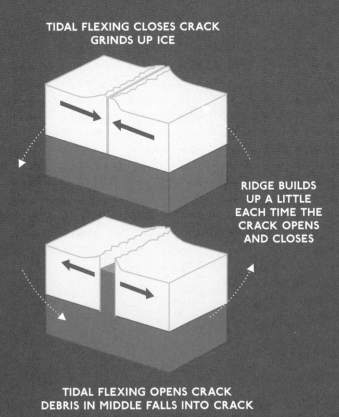

TIDAL FLEXING CLOSES CRACK GRINDS UP ICE

RIDGE BUILDS UP A LITTLE EACH TIME THE CRACK OPENS AND CLOSES

TIDAL FLEXING OPENS CRACK DEBRIS IN MIDDLE FALLS INTO CRACK

There is one final piece of evidence that completes the compelling case for the existence of liquid water beneath the surface of this frozen moon. Measurements of Europa's magnetic field and its interaction with the powerful magnetic field of its parent planet have suggested its ocean may be salt water and may be a staggering 100 kilometres (60 miles) deep. That means there is more than twice as much life-giving liquid water on this tiny moon than in all of Earth's oceans combined. This is a remarkable conclusion, especially when you consider that we've never landed on Europa, let alone burrowed through its icy shell to see what lies beneath. Fortunately and wonderfully, plans are now afoot for a joint NASA/ESA mission to launch a robotic explorer in 2020, which is being designed to land on the surface of this potential paradise moon and unlock its secrets.

If the conclusions from these various measurements of Europa are correct, and her vast salty ocean does exist, then we already know how life could, in principle, thrive there. Hidden below the ice and unable to capture the energy of the Sun, life on Europa may well mimic the ecosystems we see clustered around the hydrothermal vents in the deep oceans of our own planet; dependent on chemosynthesis rather than on photosynthesis. We don't know if Europa's ocean is too salty or too cold to support life, or if its ecosystem has remained stable for long enough for life to develop, but the discoveries we have made two-thirds of a billion kilometres away, combined with the discoveries we have made by exploring our own oceans, have created a tantalising case for this being the most likely habitable environment in our solar system ◉

VATNAJOKULL GLACIER, ICELAND

BELOW: Here at the Vatnajokull Glacier in Iceland, it seems implausible that any life form could exist in these frozen conditions, but scientists are using just such locations to disprove this theory. The ice core taken from this ice cave revealed bacteria living in the frozen sample.

It is not just the evidence for a hidden ocean that has catapulted Europa to the forefront of our search for alien life. Scientists here on Earth are beginning to rewrite our understanding of how life can withstand not just the extremes of the deep ocean but also the extreme environments on the surface of our planet. The spectacular ice caves in the Vatnajokull Glacier are a beautiful demonstration of how the laws of physics can conspire to create an awe-inspiring cathedral of ice.

This cave tunnels into the heart of the glacier, where the ice has been frozen for 1,000 years. The walls are the deepest, purest crystal blue, cut through by lines of fine silt that tell the story of the eruptions of Iceland's many volcanoes over the millennia it took for the glacier to form. But there is more to be found in these ice caves than mere beauty; they may also tell us something about what we could expect to find within the frozen ice fields of Europa.

This is what brings astro-biologists like NASA scientist Richard Hoover to locations such as these, and us alongside him. He has spent his career looking for life in the most unlikely places, and in this particular location he is interested in taking samples of the ancient ice to explore what might be lurking inside. Conventional thinking would suggest that any organisms found in ice of this age would not be living, but recent work is beginning to challenge our understanding of the boundaries of life. Once he had taken an ice core, we headed back to our freezing base in a small and isolated wooden building at the foot of the glacier. Of all the places I filmed for *Wonders of the Solar System*, our little base in the snow in the middle of Iceland in November was undoubtedly the coldest.

Richard Hoover explained the thinking behind his research ideas. 'For a long time it was thought that micro-organisms in ice were present only in the state of what is called deep antibiosis – suspended animation. It's now becoming quite clear that that isn't necessarily the case for all the micro-organisms. There may be others that are actually actively living in the ice.'

It's an incredible thought, to think that this beautiful cave could be alive, populated by living things, not frozen but truly living, dividing and reproducing. It's this prospect of finding things living in solid ice that has had an enormous impact on our ideas of where life could survive in the Solar

If life can exist here, trapped in the deep glaciers of Iceland, then is it such a leap to imagine a similar life form existing in the icy crust of Europa?

ALIENS

System. Under a microscope you can clearly see living bacteria in Hoover's sample, organisms that, it might be inferred, have been trapped in the glacier for thousands of years. You're seeing life in ice; a form of organism that Hoover believes is actually adapted to live in this frozen environment. 'We now know that some micro-organisms are capable of actually causing the ice to melt because they generate essentially anti-freeze proteins,' he explains. 'They change the temperature at which ice goes from a solid state to a liquid state, forming little tiny pockets, maybe only a few microns in diameter. If they can make a two- or three-micron diameter ball of liquid water and they have the ability to move, then that bacterium is now not in a glacier, he's in an ocean.'

If life can exist here, trapped in the deep glaciers of Iceland, then is it such a leap to imagine a similar life form existing in the icy crust of Europa? This also raises the intriguing possibility of an organic explanation for the mysterious red stains that cover the moon. The Lineae are the most striking of all of Europa's features, creating a network of colour across the entire surface of the moon. The larger formations are more than twenty kilometres (twelve miles) across, interwoven with patterns of dark and light material. This wide variety of colour could suggest the presence of microbial life, since these tell-tale colours are very reminiscent of the cyanobacteria we find on Earth. Could it really be that the ice of Europa contains viable living micro-organisms? It's a controversial idea, but it is a dizzying thought that the mysterious red stains on the surface of Europa may be the visible signs of alien life.

The question is as evocative as it is ancient. Are we alone in the Universe? Is our pale blue world the only planet amongst the billions of planets in our galaxy, amongst the billions of galaxies in the Universe, to harbour life? This is surely one of the most important and profound questions, perhaps the most important question, that we can ask. Think about what it would mean for us to have an answer. If the answer is that there is no other life in the Solar System, in our nearby star systems, perhaps throughout vast swathes of the Milky Way or even across the Universe, then how valuable would that make planet Earth? How valuable would that make us? A single island of beauty and meaning in a meaningless void. But imagine the alternative answer. What if it transpires that on every moon of every planet, where the conditions are right, life does survive and flourish. What if we discover that the Universe is teeming with life and that we are part of a vast and vital cosmic community. How might that change our behaviour? What would that tell us about the way we respond to our differences, with other species and with each other? We should then see the Earth as one village amidst a billion continents – our own local community adrift in a sea of alien life.

If knowing the answer to the question is so profoundly important, then surely striving to find the answer should be of overwhelming importance. I believe it's the most important of the great and timeless existential questions that we as a civilisation can possibly ask, because we have a chance of answering it. It would be a gross and unforgivable dereliction of our duty as civilised beings to sit tight and wonder rather than to stand up and explore.

BELOW: People crowd the streets of the Shibuya area of Japan, going about their busy lives. Few of us have time to stop and think that Earth is the only place in the Solar System where life is complex and stable enough to build a civilisation – this is what makes our planet so precious.

What we've learned from exploring such extreme places on Earth is that if there is life out there in the Solar System, it will almost certainly be simple: single-celled organisms like bacteria eking out an existence in the most hostile of environments.

One thing seems certain: the only place in the Solar System where there is life complex enough to build a civilisation is here on planet Earth. But how did that happen? What is it that makes our world so special, because, after all, everything in the Solar System shares the same genesis?

Our little corner of the galaxy was created out of nothing more than a spinning cloud of gas and dust 4.5 billion years ago. Solid worlds condensed out of the swirling mists, but each world was radically different. Across the Solar System there are worlds that erupt with volcanoes of sulphur and others with geysers of ice; there are worlds with rich atmospheres and swirling storms, and there are moons with surfaces sculpted from ice that conceal oceans of liquid water. But amidst all the wonders, there's only one world where the laws of physics have conspired to combine all these features in one place.

Only on Earth are the temperatures and atmospheric pressure just right to allow oceans of liquid water to exist on the surface of the planet. Earth is big enough to have retained its molten core that not only powers geysers and volcanoes,

but also produces our magnetic field that fends off the solar wind and protects our thick, nurturing atmosphere.

It is the combination of all these wonders in one place that allowed life to begin and to get a foothold here on Earth. Yet to allow that life to evolve into such complex creatures as ourselves requires one more ingredient, and that is time, deep time – the vast and sweeping vistas of time over which mountains rise and fall, planets are formed and stars live and die. When all is said and done, it is perhaps this that makes the Earth so rare and so precious in the cosmos, because it has been stable enough for long enough for life to evolve into such magnificent complexity.

Life on Earth today is the result of hundreds of millions of years of stability, and the most remarkable, complex and wonderful component in this priceless and possibly uniquely sophisticated and interconnected ecosystem is us, humankind, a species that has developed to the point where we can bend and shape and change the world around us. We have even left our home planet behind to begin exploring our cosmic surroundings. We are powerful, and with that power comes great responsibility, for we are now the custodians of Earth. Through our evolved intelligence, we have the capability to protect, damage or destroy it as we chose. And it is a choice – a choice that is best informed by perspective ◉

You could take the view that our exploration of the Universe has made us somehow insignificant; one tiny planet around one star amongst hundreds of billions. But I don't take that view, because we've discovered that it takes the rarest combination of chance and the laws of Nature to produce a planet that can support a civilisation, that most magnificent structure that allows us to explore and understand the Universe. That's why, for me, our civilisation is the wonder of the Solar System, and if you were to be looking at the Earth from outside the Solar System that much would be obvious. We have written the evidence of our existence onto the surface of our planet. Our civilisation has become a beacon that identifies our planet as home to life.

'We shall not cease from exploration.
And the end of all our exploring
Will be to arrive where we started
And know the place for the first time.'

— T. S. Eliot

RIGHT: Having begun to explore our cosmic surroundings, we are more aware than ever before of our fragility and our responsibility to protect planet Earth in the future.

INDEX

Entries in *italics* denote photographs and diagrams

greenhouse gases/effect 12, 126, 127, 134, 136, *136–7*
gypsum 223

H

Hale Bopp comet *110–11*
Hawaii *110–11*, 168, *168*, 174, 178, 184, *184*, 185
heliopause 58
heliosphere 50–1, 55
Herschel, Caroline 98
Herschel, John 30
Herschel, Sir William 26, 30, 98, 120, 185, 210–11
Hertzsprung-Russell diagram 62, *62–3*, 65
Hilderbrand, Dr Alan 130, 131, *131*
Hoover, Richard 236, 239
Hubble Space Telescope 9, 24, 25, 55, 80, *118–19*, *123*
human make-up 212
Huygens, Christopher 24, 88
Hyperion 94, *95*

I

Iapetus 88, 94, *95*
Iceland 55
    Great Geysir, Haukadalur Valley, Iceland 100
    icebergs in 90, *90–1*
    landscape of *98*, 99
    Strokkur Geyser, Haukadalur Valley, Iceland 100
    Vatnajokull glacier 236, *236–9*, 239
Iguaçu Falls, border of Brazil and Argentina 37, *37*, 40, *40*, *41*
India:
    Deccan Traps 178–9, *178*
    Ganges river 23, *23*, 45, *45*
    monsoon in 40
    solar eclipse in *20*, 23, *23*, 44–5, *44–5*
    Varanasi *20*, 23, *23*, 44–5, *44–5*
Infrared Telescope Facility, NASA *110–11*
interstellar wind 58, 60
Io 24, 55, 192, 194–5, *194*, *195*, 198–9, *198*, *199*, 200, 234

J

Japan *240*
Jupiter 9, 27, 86, 88, *181*
    asteroids, gravitational influence over passing 183, 184, 185, 186, 188, 189
    atmosphere 127

eclipses 24
ethereal planet 182, *182*, 183, *183*
gravitational influence over solar system 180, 182–3, 184, 185, 186, 188, 189, 198–9, *198*, *199*
Great Red Spot *8–9*, *146*, 147, *182*
Late Heavy Bombardment and 9, 110
magnetic field *55*
moons 24, *25*, 55, 94, 192, *192*, *193*, 198–9, *198*, *199*
north pole *182*
orbit of Sun 71
rotations of *183*
Saturn and 108
size of 180, 182
south pole *182*
surface temperature 128
Voyager missions to 58
weather *8–9*, *146*, 147

K

Kairouan, great mosque of 69, *69*
Kansas, USA *145*
Keck Observatory *110–11*, 228–9
Kennedy, John F. 15, 16
Kennedy Space Centre, Florida *14–15*, 16, 58, *59*, 108
Kilauea, Hawaii 168, *168*
Kuiper, Gerard 9, 151
Kuiper comet belt 9, 151

L

Lake Eyak, Alaska 158
Lake Missoula 215–16
Lake Thorisvatn, Iceland 55
Late Heavy Bombardment 9, 106, *107*, 108, *109*, 110
lava lakes 194, *194*, *195*, *196–7*, 197, *198*, 199, *199*, 200
Levy, David 50
life, what is? 212
light:
    colours and wavelengths of 43
    speed of 24
'limb darkening' 151
linear momentum 79
Lowell, Percival 211, 218, 229
Luna 1 8

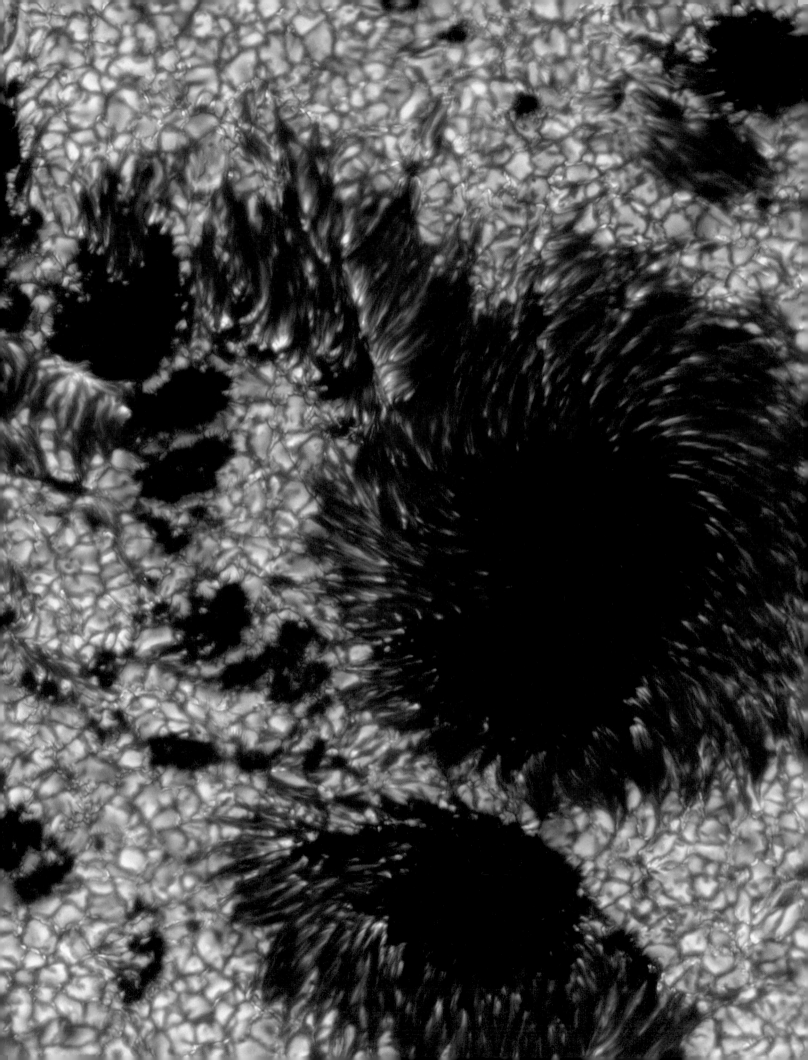

M

Magellan probe 134, *177*, 178
magnetosphere 50
Mallory, George 16
Mariner 4 probe 211
Mariner 9 probe 166, 169
Mariner 10 probe 132, *133*
Mars:
　Arsia Mons 224, *224*
　Ascraeus Mons *224*
　atmosphere 127, 139, 141, 142–3
　caves on 224
　clouds on 166
　Earth, similarities with 166, *167*
　eclipses 24–5
　Endurance Crater 222–3, *223*
　extraterrestrial life on 211, 213, 218–29, 230
　familiar world 166, *166*, *167*
　Gusev Crater 27, *27*
　gypsum on 223
　'Heat Shield Rock' 139
　landscape formed by water 166, *167*
　loses heat and becomes inactive 176, 177
　meteorites on 139, *139*
　methane on 228–9
　minerals 222, *222*, 223, *223*
　moons 24, *25*
　movement of in Earth's sky 74, *75*
　nuclear winter on 12
　Olympus Mons 169, *169*, 175, *224*
　orbit of Sun 71
　origins of 169
　Pavonis Mons *224*
　probes to 24–5, 27, *27*, 138, 139, *139*, *140–1*, 141, 166, *219*, 222–3, *223*
　retrograde motion 74, *75*, 76, *77*
　scars on 220, *220*, *221*
　size of 94
　skies 166, *166*
　sunset/sunrise on 27, *27*
　surface temperature 129, 166, 179
　Tharsis 169
　Tharsis Montes *224*
　unexplored subterranea 224, *224*, 227
　Valles Marineris 164, *164–5*, 166, 169
　Victoria Crater *219*
　volcanoes of 169, *169*, 175, 224
　water on 211, 213, 220, *220*, *221*, 223, *223*, 234
Mars Antenna, NASA Deep Space Network, California 58, *210–11*
Mars Exploration Rover:
　Opportunity 24–5, 139, *139*, 166, *219*, 222–3, *223*
　Spirit 27, *27*, *138*, 139, 166, 222
Mars Express satellite 166
Mars Odyssey satellite 166, 224, *224*

Mars Reconnaissance satellite 166
Mars 3 probe 139
Matanuska glacier, Alaska *154–5*, 155
Mauas, Pablo 40, *40*, 41
Mauna Kea, Hawaii *110–11*, 168, 178
Mercury 27, 94, 231
　asteroids and comets, bombardment by 132, *133*
　craters 132, *132*, *133*
　engulfed by the Sun 65
　loss of atmosphere 127, *127*, 132, *133*, *149*
　orbit of Sun 71, 104
　origin of 169
　temperatures on *127*, 128
Messenger probe 132, *133*
meteorites *184 see also* asteroids
methane 228–9, *228*, 229
Milky Way 31, 60
Mimas 104–5, 108
Minas Basin, Bay of Fundy, Nova Scotia, Canada *190*, *191*
Molecular Cloud Barnard 68 30, *31*
Moon 8
　Earth's tides and 190–1, *190*, *191*, 192
　Late Heavy Bombardment and 106, 108, *109*
　Mare Imbrium 108
　size of 9
　stabilises Earth's seasons 9
Moore, Patrick 182
Murray, Professor Carl 89

N

Namib Desert, Namibia 126, *126*, 139, *140*
NASA 25, 58, 88, *110–11*, 119, 132, *133*, 134, 139, 162, *164–5*, 166, 169, *177*, 178, 182, 192, *210–11*, 210, 211, 224, 228–9, 230, 232, 233, 235, 236
NATO 116
Neptune 9, 26, 86, 151
　atmosphere 127
　discovery of 120, 121, *121*
　eclipses 24
　Late Heavy Bombardment and 9, 108
　moons 24
　orbit of Sun 71
　surface temperature 129
　Voyager missions to 58
Newton, Sir Isaac 24, 25, 103, 104, 120, 121, 176, 192
NGC 1672 (galaxy) *10–11*
North Star (Polaris) 74
nuclear fusion 34–5
nuclear winter 12

250

U

V

W

Y

Z

PICTURE CREDITS

All pictures are copyright the BBC except:

18, 50, 62, 72, 86, 128, 142, 172, 200 Nathalie Lees ©
HarperCollins; 8–9 Corbis; 10–11 NASA-Hubble Heritage/
digital/Science Faction/Corbis; 13 NASA/digital/Science
Faction/Corbis; 28 Diego Giudice/Corbis; 47 Roger Ressmeyer
with Jay Pasachoff/Williams College/Science Faction/
Corbis; 56–57 Arctic-Images/Corbis; 70 Peter Richardson/
Robert Harding World Imagery/Corbis; 110 David Nunuk/
Science Photo Library; 138 HO/Reuters/Corbis; 145 Jim
Reed/Corbis; 174 Corbis; 181 Bettmann/Corbis; 209 National
Undersea Research Program/NOAA/Science Photo Library;
228 Sinclair Stammers/Science Photo Library; 240 TWPhoto/
Corbis; 248 Scharmer et al, Royal Swedish Academy of
Sciences/Science Photo Library; 253 NASA/JPL/SSI/Science
Photo Library; 14, 17, 25 (both), 31, 35, 38 (both), 48, 53, 58,
59, 60 (top), 61 (top), 75, 80, 81, 82, 89 (left), 95 (top 2), 100
(bottom), 108, 109, 118, 121, 122, 123, 133 (all), 134, 135, 139, 140
(top), 146, 147, 148, 150, 153, 156, 159, 164, 169 (both), 170, 177
(both), 179 (both), 182, 183, 185 (both), 188 (top), 190, 191, 192,
193, 198, 219, 224 (left), 231, 232, 233 (all), 243 NASA; 32-33
(bottom), 39, 60 (bottom), 69 (both), 85, 92, 97, 100 (top), 157
Prime Focus/BBC; 41, 42, 54 (bottom), 90 (top), 107, 125, 136,
154, 236 Brian Cox

P248: Sunspots are cooler regions
of the Sun's surface that appear
dark against their brighter, hotter
surroundings.

P253: Saturn and its rings, as seen
by the wide-angled camera on
NASA's Cassini spacecraft.

ACKNOWLEDGEMENTS

In writing this book we are indebted to the hard work and creativity of all those who were involved in the BBC television production of *Wonders of the Solar System*. Danielle Peck who lead the team with great passion and drive, Gideon Bradshaw, Michael Lachmann, Chris Holt and Paul Olding who produced and directed such beautiful films.

We'd also like to thank, Rebecca Edwards, Diana Ellis-Hill, Tom Ranson, Ben Finney, Laura Mulholland, Kevin White, George McMillan, Paul Jenkins, Simon Farmer, Chris Titus King, Freddie Claire, Darren Jonusas, Gerard Evans, Martin Johnson, Louise Salkow, Lee Sutton, Laura Davey, Rebecca Lavender, Alison Castle and David Pembrey, Sheridan Tongue, Donna Dixon, Julie Wilkinson, Sara Revell, Anna Charlton, Louise Farley, Nicola Kingham and the team at Prime Focus London.

We'd also like to thank Dr Andy Pilkington for his work on the book and Professor Jeff Forshaw and Professor John Zarnecki for the generous time and thought they gave to the project.

Brian would also like to thank The University of Manchester and The Royal Society for allowing him the time to make *Wonders*. He is also particularly indebted to Professor Alan Gilbert (11 September 1944 – 27 July 2010), the Inaugural President and Vice Chancellor of The University of Manchester, who knew the true value of universities and encouraged his institution and academics to make a difference to society far beyond the ivory towers.